T0338611

Smart Manufacturing Technologies for Industry 4.0

This book addresses issues related to the integration of digital evolutionary technologies and provides solutions to various challenges encountered during the implementation process. With real-time case studies, the book explains the smart technologies available and their operational applications and benefits in the manufacturing sector.

Smart Manufacturing Technologies for Industry 4.0: Integration, Benefits, and Operational Activities assists in the understanding of the shifting paradigm in the manufacturing sector towards smart manufacturing and spotlights these technologies and the effects they are having on existing industries. It showcases Industry 4.0 as a promising research area in its infancy and offers insights into the role smart technologies are playing now and into the future. The book focuses on smart technologies' rudiments, implementation, and integration for organizational development and offers insights on how to achieve resiliency through and because of these technologies.

This book presents real-time implementation discussions along with case studies that emphasize benefits and operational activities for engineers and managers. It's also a very useful book for technology developers, academicians, data scientists, industrial engineers, researchers, and students interested in uncovering the latest innovations in a field that seeks current research on products and services.

Advances in Intelligent Decision-Making, Systems Engineering, and Project Management

This new book series will report the latest research and developments in the field of information technology, engineering and manufacturing, construction, consulting, healthcare, military applications, production, networks, traffic management, crisis response, human interfaces, and other related and applied fields. It will cover all project types, such as organizational development, strategy, product development, engineer-to-order manufacturing, infrastructure and systems delivery, and industries and industry-sectors where projects take place, such as professional services, and the public sector including international development and cooperation etc. This new series will publish research on all fields of information technology, engineering, and manufacturing including the growth and testing of new computational methods, the management and analysis of different types of data, and the implementation of novel engineering applications in all areas of information technology and engineering. It will also publish on inventive treatment methodologies, diagnosis tools and techniques, and the best practices for managers, practitioners, and consultants in a wide range of organizations and fields including police, defense, procurement, communications, transport, management, electrical, electronic, aerospace, requirements.

Blockchain Technology for Data Privacy Management
*Edited by Sudhir Kumar Sharma, Bharat Bhushan, Aditya Khamparia,
Parma Nand Astya, and Narayan C. Debnath*

Smart Sensor Networks Using AI for Industry 4.0
Applications and New Opportunities
*Edited by Soumya Ranjan Nayak, Biswa Mohan Sahoo, Muthukumaran Malarvel,
and Jibitesh Mishra*

Hybrid Intelligence for Smart Grid Systems
*Edited by Seelam VSV Prabhu Deva Kumar, Shyam Akashe, Hee-Je Kim,
and Chinmay Chakrabarty*

Machine Learning-Based Fault Diagnosis for Industrial Engineering Systems
Rui Yang and Maiying Zhong

For more information about this series, please visit: www.routledge.com/Advances-in-Intelligent-Decision-Making-Systems-Engineering-and-Project-Management/book-series/CRCAIDMSEPM

Smart Manufacturing Technologies for Industry 4.0

Integration, Benefits, and Operational Activities

Edited by Jayakrishna Kandasamy, Kamalakanta Muduli,
V.P. Kommula, and Purushottam Meena

CRC Press
Taylor & Francis Group
Boca Raton London New York

CRC Press is an imprint of the
Taylor & Francis Group, an **informa** business

First edition published 2023
by CRC Press
6000 Broken Sound Parkway NW, Suite 300, Boca Raton, FL 33487–2742

and by CRC Press
4 Park Square, Milton Park, Abingdon, Oxon, OX14 4RN

CRC Press is an imprint of Taylor & Francis Group, LLC

ISBN: 978-1-032-03308-2 (hbk)
ISBN: 978-1-032-03309-9 (pbk)
ISBN: 978-1-003-18667-0 (ebk)

DOI: 10.1201/9781003186670

Typeset in Times New Roman
by Apex CoVantage, LLC

Contents

Editors vii

Contributors ix

Abbreviations xi

Chapter 1 Organizational Transformation towards Industry 4.0 Technologies 1

Manavalan E and Jayakrishna Kandasamy

Chapter 2 The Autonomy of Autonomous Robots 9

Raunika Anand, Arijeet Nath, and S. Aravind Raj

Chapter 3 Smart Technologies for Industry 4.0 and Its Future 19

P. Ben Rajesh and A. John Rajan

Chapter 4 Digital Twin–Based Smart Manufacturing—Concept and Applications 31

Vishal Ashok Wankhede, Rohit Agrawal, and S. Vinodh

Chapter 5 Applications of Augmented and Virtual Reality in Contemporary
Manufacturing Organisations 37

Santosh Kumar, Vijaya Kumar, and Sakthi Balan Ganapathy

Chapter 6 Cloud-Based Manufacturing Service Selection Using Simulation
Approaches 49

Vaibhav S. Narwane, Irfan Siddavatam, and Rakesh D. Raut

Chapter 7 Go Unsupervised via Artificial Intelligence 57

Amit Vishwakarma, G.S. Dangayach, M.L. Meena, and Sumit Gupta

Chapter 8 Integration of Cyber-Physical Systems for Flexible Systems 65

*Thirupathi Samala, Vijaya Kumar Manupati,
Bethalam Brahma Sai Nikhilesh, and Jose Machado*

Chapter 9 Wearables to Improve Efficiency, Productivity, and Safety of
Operations 75

K. Balamurugan, T.P. Latchoumi, and T.P. Ezhilarasi

Chapter 10 Analysis of Factors Influencing Cloud Computing Adoption in
Industry 4.0-Based Advanced Manufacturing Systems 91

Anilkumar Malaga and S. Vinodh

Chapter 11 Liberating 3D Printing for a New Normal in Manufacturing 103

*Tarun Kataray, Mayank Mishra, Vaishnav Madhavadas,
Deepesh Padala, Bhawana Choudhary, Swati Kashyap,
Utkarsh Chadha, and S. Aravind Raj*

Index 115

Editors

Dr. Jayakrishna Kandasamy is an associate professor in the School of Mechanical Engineering at the Vellore Institute of Technology University, India. Dr. Jayakrishna's research is focused on the design and management of manufacturing systems and supply chains to enhance efficiency, productivity, and sustainability performance. His more recent research is in the area of developing tools and techniques to enable value creation through sustainable manufacturing, including methods to facilitate more sustainable product design for closed-loop material flow in industrial symbiotic setups and developing sustainable products using hybrid biocomposites. He has mentored doctoral, graduate, and undergraduate students, which has so far led to 72 journal publications in leading SCI/SCOPUS indexed journals, 29 book chapters, 96 refereed conference proceedings, 4 authored books, and 11 edited books in CRC Press/Springer series. Dr. Jayakrishna's team has received numerous awards in recognition for the quality of the work that has been produced. He teaches undergraduate and graduate courses in the manufacturing and industrial systems area, and his initiatives to improve teaching effectiveness have been recognized through national awards. He was also awarded the Global Engineering Education Award from Industrial Engineering and Operations Management (IEOM) Society International, USA, in 2021; Institution of Engineers (India)—Young Engineer Award in 2019; Distinguished Researcher Award in the field of sustainable systems engineering in 2019 by the International Institute of Organized Research; and Best Faculty Researcher Award for the year(s) 2016–2021 consecutively. He is the coordinator—Circular Economy Club, VIT University; academic editor—Mathematical Problems in Engineering, Wiley-Hindawi Publications; editorial board member—*Journal of Operational Research for Engineering Management Studies* (JOREMS); area editor—*Operations Management Research*; associate editor—*Circular Economy, Frontiers in Sustainability*; and book series editor—Industrial Engineering, Systems, and Management, CRC Press.

Dr. Kamalakanta Muduli is presently working as associate professor in the Department of Mechanical Engineering, Papua New Guinea University of Technology, Lae, Morobe Province, Papua New Guinea. He obtained a PhD from the School of Mechanical Sciences, IIT Bhubaneswar, Orissa, India. He obtained a master's in industrial engineering. Dr. Muduli has over 16 years of academic experience in universities in India and Papua New Guinea. Dr. Muduli is a recipient of the ERASMUS+ KA107 award provided by the European Union. He has published 64 papers in peer-reviewed international journals, most of which are indexed in Clarivate analytics and Scopus and listed in ABDC, and more than 31 papers in national and international conferences. He has also guest edited a few special issues in journals and books approved for publication by Springer, Taylor & Francis Group, MDPI, CRC Press, Wiley Scrivener, and Apple Academic Press. Dr. Muduli also has guided three PhD students. His current research interests include materials science, manufacturing, sustainable supply chain management, and Industry 4.0 applications in operations and supply chain management. Dr. Muduli is a fellow of the Institution of Engineers India. He is also a senior member of the Indian Institution of Industrial Engineering and a member of ASME.

Dr. V.P. Kommula is currently working as a professor (full) at the Department of Mechanical Engineering, University of Botswana. He has more than 22 years of teaching experience and served in various positions at different universities in countries like India, Malaysia, Republic of South Africa, and Botswana. He teaches courses related to engineering design, materials, manufacturing, industrial engineering, engineering management, and CAD/CAM/CIM at the undergraduate level. He teaches and supervises postgraduate students in engineering management and manufacturing streams. His active research areas are natural fiber composite materials, lean manufacturing, productivity improvement, and Lean Six Sigma. He has a specific interest in engineering education.

Dr. Purushottam Meena is currently working as Supply Chain Management faculty member in the School of Business at the University of Charleston, SC, USA. he previously worked as an associate professor of operations management in the School of Management at the New York Institute of Technology. He holds a PhD in industrial engineering and management. His primary areas of research interest include big data analytics, blockchain applications in supply chain, reshoring, sustainable operations, risk management, omnichannel supply chain, supply chain performance, buyer-supplier relationships, and analytics. His PhD dissertation won the 2012 Emerald/EFMD Outstanding Doctoral Research Award and was a finalist for the 2013 CSCMP Doctoral Dissertation Award. Dr. Meena was conferred the 2021 and 2018 Distinguished NYIT—SoM Faculty Scholarship Award. He has authored more than 50 peer-reviewed journal and conference articles. His research work has appeared in leading journals, including *Transportation Research Part E, International Journal of Production Economics, Journal of Retailing and Consumers Research, International Journal of Advanced Manufacturing Technology, Journal of Business & Industrial Marketing*, and *Industrial Management and Data Systems*. Dr. Meena has consulted with several companies in the United States and India. He has delivered keynote talks at different international conferences and served on the editorial boards of more than ten journals in the operations and supply chain field.

Contributors

Rohit Agrawal
Operations Management & Quantitative
    Techniques Division
Indian Institute of Management
Bodh Gaya, Bihar, India

Raunika Anand
Department of Manufacturing Engineering
School of Mechanical Engineering
Vellore Institute of Technology
Vellore, Tamil Nadu, India

K. Balamurugan
Department of Mechanical Engineering
Vignan Foundation for Science
Technology and Research
Tenali, Andhra Pradesh, India

Utkarsh Chadha
Department of Manufacturing Engineering
School of Mechanical Engineering
Vellore Institute of Technology
Vellore, Tamil Nadu, India

Bhawana Choudhary
School of Computer Science and Engineering
Vellore Institute of Technology
Vellore, Tamil Nadu, India

G.S. Dangayach
Department of Mechanical Engineering
Malaviya National Institute of Technology
Jaipur, India

Manavalan E
Delivery Manager, Kinaxis Inc
Chennai, Tamil Nadu, India

T.P. Ezhilarasi
Department of Computer Science and Engineering
Raak College of Engineering & Technology
Tamil Nadu, India

Sakthi Balan Ganapathy
Department of Manufacturing Engineering
School of Mechanical Engineering
Vellore Institute of Technology (VIT)
Vellore, Tamil Nadu, India

Sumit Gupta
Department of Mechanical Engineering
Amity School of Engineering and Technology
Amity University
Noida, Uttar Pradesh, India

Jayakrishna Kandasamy
School of Mechanical Engineering
Vellore Institute of Technology
Vellore, Tamil Nadu, India

Swati Kashyap
School of Electronics Engineering
Vellore Institute of Technology
Amaravati, Andhra Pradesh, India

Tarun Kataray
School of Mechanical Engineering
Vellore Institute of Technology
Vellore, Tamil Nadu, India

Santosh Kumar
Ramakrishna College of Engineering
Samayapuram, Tiruchirappalli
Tamil Nadu, India

Vijaya Kumar
Ramakrishna College of Engineering
Samayapuram, Tiruchirappalli
Tamil Nadu, India

T.P. Latchoumi
Department of Computer Science and
    Engineering
SRM Institute of Science and Technology
Chennai, India

Jose Machado
Department of Mechanical Engineering,
    School of Engineering
University of Minho, Portugal

Vaishnav Madhavadas
Department of Manufacturing Engineering
School of Mechanical Engineering
Vellore Institute of Technology
Vellore, Tamil Nadu, India

Anilkumar Malaga
Department of Production Engineering
National Institute of Technology
Tiruchirappalli, Tamil Nadu, India

Vijaya Kumar Manupati
Operations and Supply Chain Division
National Institute of Industrial Engineering
Mumbai, India

M.L. Meena
Department of Mechanical Engineering
Malaviya National Institute of Technology
Jaipur, India

Mayank Mishra
School of Mechanical Engineering
Vellore Institute of Technology
Vellore, Tamil Nadu, India

Vaibhav S. Narwane
K J Somaiya College of Engineering
Mumbai, India

Arijeet Nath
Department of Manufacturing Engineering
School of Mechanical Engineering
Vellore Institute of Technology
Vellore, Tamil Nadu, India

Bethalam Brahma Sai Nikhilesh
Department of Mechanical Engineering
National Institute of Technology
Warangal, Andhra Pradesh, India

Deepesh Padala
School of Electronics Engineering
Vellore Institute of Technology
Vellore, Tamil Nadu, India

S. Aravind Raj
Department of Manufacturing Engineering
School of Mechanical Engineering
Vellore Institute of Technology
Vellore, Tamil Nadu, India

A. John Rajan
Department of Manufacturing Engineering
School of Mechanical Engineering
Vellore Institute of Technology
Vellore, Tamil Nadu, India

P. Ben Rajesh
Department of Mechanical Engineering
Sathyabama Institute of Science and
    Technology
Chennai, Tamil Nadu, India

Rakesh D. Raut
Operations & Supply Chain Management
    Division
National Institute of Industrial Engineering
Mumbai, India

Thirupathi Samala
Department of Mechanical Engineering
National Institute of Technology
Warangal, Andhra Pradesh, India

Irfan Siddavatam
Department of Mechanical Engineering
K J Somaiya College of Engineering
Mumbai, India

S. Vinodh
Department of Production Engineering
National Institute of Technology
Tiruchirappalli, Tamil Nadu, India

Amit Vishwakarma
Department of Mechanical Engineering
Malaviya National Institute of Technology
Jaipur, India

Vishal Ashok Wankhede
Department of Production Engineering
National Institute of Technology
Tiruchirappalli, Tamil Nadu, India

Abbreviations

AI   Artificial Intelligence
CPS   Cyber Physical System
ICT   Information and Communication Technology
IIoT Industrial Internet of Things
IoT   Internet of Things
IS   Information Systems
IT   Information Technology

1 Organizational Transformation towards Industry 4.0 Technologies

Manavalan E and Jayakrishna Kandasamy

CONTENTS

1.1   Introduction..1
1.2   Overview of Digital Supply Chain...1
1.3   Supply Chain Challenges..2
1.4   Digital Transformation with Industry 4.0...2
1.5   Influence of Industry 4.0 in Organizations..3
1.6   Next Wave of Industry 4.0..3
1.7   Building Blocks of Industry 4.0...4
1.8   Roadmap for Industry 4.0 Transformation...5
1.9   Theoretical and Practical Implications...6
1.10 Conclusion..6
1.11 Limitations and Future Research Directions..7
References...7

1.1   INTRODUCTION

The supply chain comprises vendors, producers, wholesalers, retailers, and end clients, and it is intended to synchronize demand and supply (Pereira and Frazzon, 2021). Globalization in the current scenario is confronted by the constant growth of global demand in capital and consumer goods by persistently verifying overall interest in the social, environmental, and economic aspects (Habib *et al.*, 2020).

1.2   OVERVIEW OF DIGITAL SUPPLY CHAIN

The synchronized value creation process from raw material purchase to end-client usage can be effectively accomplished by better management of the supply chain. The benefits of information systems are increased efficiency, increased productivity, reduced time, cost reduction, zero errors, and optimized inventory (Nunez-Merino *et al.*, 2020; Asamoah *et al.*, 2021). Digital disruption has become the order of the day. The convergence of information systems has become more complicated and dynamic, which drives industrial organizations to invest in smart manufacturing (Ante, 2021). Organizations must embrace changing technologies to cope with a shorter product life cycle and rapid environmental changes. Information systems (ISs) attempt to integrate communication between people and technology. An IS, including an enterprise resource planning (ERP) system, provides a seamless user experience in real time that is intuitive enough to make informed decisions and support effective management of the overall operations of the organization (Asamoah *et al.*, 2021).

DOI: 10.1201/9781003186670-1

1.3   SUPPLY CHAIN CHALLENGES

In the present dynamic supply chain environment, an organization faces many challenges in the manufacturing space, such as global competitiveness, lack of adaptability, and go-to-market time (Bressanelli *et al.*, 2019; Yadav *et al.*, 2020). The common challenge faced by manufacturers is to forecast the right demand versus supply and to reduce manufacturing lead time (Abolghasemi *et al.*, 2020). Consequently, it becomes necessary for manufactures to equip themselves with the right resources and processes along with technology to provide revolutionary products and world-class services. As illustrated in Figure 1.1, factors that influence uncertainty in supply chain management are global competition, lack of adaptability, and delayed entry into the market.

IoT technology overcomes these challenges, which significantly transforms the supply chain industry (de Vass *et al.*, 2021). For instance, this technology can be leveraged to track the consignment location and speed of a vehicle so that users are alerted about late deliveries. IoT technology can be deployed to monitor the condition of equipment from a remote location (Jamil *et al.*, 2020). Temperature-sensitive products can be monitored with sensors, and the data can be communicated through the internet. The combination of internet, wireless, predictive analytics, and cloud technologies can change the entire supply chain operation and bring more value to the organization (Ivanov and Dolgui, 2020). Using IoT, wastage can be minimized, as it monitors the condition of perishable products and sends the status to the stakeholders of the supply chain (Mastos *et al.*, 2021).

1.4   DIGITAL TRANSFORMATION WITH INDUSTRY 4.0

The phenomenon of Industry 4.0 was first proposed by Germany to take strategic initiatives in 2011. Initially, the perception of Industry 4.0 was twofold. First, it was expected that the industrial revolution would have a huge impact on the economy. Second, Industry 4.0 promises high operational effectiveness and gives a path to adopt new business models (Fatorachian and Kazemi, 2021). The German government is investing in Industry 4.0 and promoting this revolution in response to the European debt crisis. They are the global leader in the manufacturing, automotive, electronics, and sports equipment industries (Gupta *et al.*, 2020). The impact of Industry 4.0 in manufacturing systems is tremendous, and it is the key to transforming machine-driven manufacturing to digital-driven manufacturing (Oztemel and Gursev, 2018).

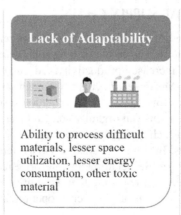

FIGURE 1.1   Key challenges for supply chain organizations.

Industry 4.0 is defined as

> a collective term for technologies and concepts of value-chain organizations that connects people, things, and systems and it creates dynamic, self-organizing, real-time optimized value-added connections within and across companies. These can be optimized according to different criteria such as costs, availability, and consumption of resources.
>
> (Gilchrist, 2016)

Industry 4.0 is a blend of digital technologies, which pushes industrial production to the next level. Industry 4.0 will alter the complete production, operations, and maintenance of products and services through interconnected components, machines, and humans. With the influence of Industry 4.0, industrial production systems are expected to perform 30% faster than before and 25% more efficiently (Boston Consulting Group, 2015). The fundamental platform for the success of this technology is the revolution in technologies that consist of cyber-physical systems (CPS), the Internet of Things (IoT), and artificial intelligence (AI). It is flexible and intelligent to interconnect machines that enable the making of customized products (Hahn, 2020).

1.5 INFLUENCE OF INDUSTRY 4.0 IN ORGANIZATIONS

Circularity focuses on the three principles: minimize waste, keep parts and goods in use, and utilize natural energy. To achieve these principles requires a study of existing processes and systems. Industry 4.0 helps to improve existing processes and systems with a focus on economic and environmental practices. The development in Industry 4.0 gives tremendous opportunities to realize a circular supply chain (Birkel and Muller, 2020). Stock and Seliger (2016) provided an overview of opportunities available for Circular Supply Chain in Industry 4.0.

With the help of Industry 4.0 technologies and retrofitting approaches, organizations can extend equipment life by studying the eco-friendly characteristics of sustainable product manufacturing. A viable option for small and medium-sized enterprises is to re-use equipment, as they may not be able to afford the significant capital to procure new manufacturing equipment (Lins and Oliveira, 2020). Organizations should leverage opportunities to enable ways to re-use and re-manufacture products to achieve higher-level throughput.

1.6 NEXT WAVE OF INDUSTRY 4.0

Industry 4.0 can be realized only when organizations are ready to transform to digital technology. With the influence of Industry 4.0, traditional supply chains have great potential to transform a highly efficient digital supply chain by smartly connecting product development, procurement, manufacturing, logistics, suppliers, customers, and service (Garrido-Hidalgo *et al.*, 2019). The entire ecosystem will be benefited if Industry 4.0 can go one step further in accommodating the sustainability of the digital supply chain. Figure 1.2 visualizes the extensive interconnection of

FIGURE 1.2 Sustainable supply chain network with future Industry 4.0.

components, machines, systems, processes, and various stakeholders of the supply chain to form a digital supply chain network with future Industry 4.0.

Disruptive business models can offer smart products and services to serve customers in a complete digital ecosystem. The digital cybernetwork is the focal point, which shares information across the supply chain, mainly distinguished by taking decentralized actions from a remote location within the tightly integrated platform (MacCarthy and Ivanov, 2022).

1.7   BUILDING BLOCKS OF INDUSTRY 4.0

The manufacturing sector is a central focal point for implementing new technologies (Erol *et al.*, 2016). Digital re-invention in the industrial revolution is set to redefine every entity in the manufacturing value chain. Digital data, connectivity, automation, and mobility are the digital disruptions happening with Industry 4.0. A list of technological trends is shown in Table 1.1, and these are instrumental in contributing to Industry 4.0 growth.

TABLE 1.1
Industry 4.0 Technologies That Help Digital Transformation

S. No	Technology	Description	Source
1	Big data–based quality management	Algorithms based on historical data detect quality concerns and decrease product failures.	Sahal *et al.*, 2020
2	Cybersecurity	Cybersecurity measures take high priority as they recognize the new vulnerabilities and challenges that interlink industrial management processes and systems digitally.	Corallo *et al.*, 2020
3	Auto-coordinated Production	Automatically coordinated machines optimize their utilization and output.	Zhang *et al.*, 2017
4	Smart supply linkage	Monitoring the supply network allows for better supply judgments.	Radziwon *et al.*, 2014
5	Self-transported vehicles	Fully automated transportation systems are used logically within the industry.	Hofmann and Rusch, 2017
6	Augmented job, maintenance, and repair operations	Emerging methods, which facilitate maintenance guidance, remote support, and service.	Hernandez-de-Menendez *et al.*, 2020
7	Lean modernization	Lean automation brings flexibility and eliminates redundant manufacturing efforts for a whole range of machines.	Kolberg and Zühlke, 2015
8	Additive manufacturing	Products are created using 3D printers that reduce product development costs.	Huang *et al.*, 2013
9	Manufacturing operations simulation	Simulation helps in optimizing the assembly line using optimization applications.	Florescu and Barabas, 2020
10	Maintenance cloud service	Manufacturers offer maintenance services rather than a product and build private clouds to save appropriate manufacturing details and processing.	Yue *et al.*, 2015
11	Flat and hierarchical system integration	Integrate cross-functional departments as well as providing seamless supply chain coordination among SC partners by automating the process wherever necessary.	Shamim *et al.*, 2017
12	Robot-aided manufacturing	Flexible, intelligent robots perform operations such as assembly and packaging independently.	Wang *et al.*, 2017
13	IIoT	This technology is an essential part of Industry 4.0 equipped to interact and communicate with the smart factory and supply chain.	Wu *et al.*, 2020
14	Intuitive predictive maintenance	Remote monitoring of equipment permits repair before breakdown.	Wang *et al.*, 2016

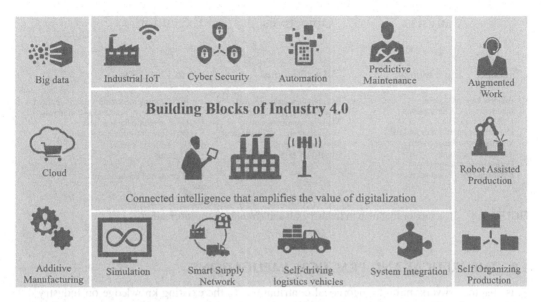

FIGURE 1.3 Industry 4.0 technologies.

Industry 4.0 technologies in manufacturing result in increased speed to market, accuracy in the product, customized output as required by the customer, and improved overall efficiency. The advancements in technology, coupled with connected intelligence that amplifies the value of digitalization, are the foundations of Industry 4.0, as illustrated in Figure 1.3. It is estimated that Industry 4.0 will redefine businesses for the next decade. The IoT influenced Industry 4.0 and provides greater efficiencies in production with fully integrated, automated, and optimized processes (Manavalan and Jayakrishna, 2019).

1.8 ROADMAP FOR INDUSTRY 4.0 TRANSFORMATION

The first phase in the roadmap to Industry 4.0 and circularity transformation is the Analysis phase, as illustrated in Figure 1.4. In this phase, the key managerial requirements are analyzed to find the knowledge level of the stakeholders, and it is important to create awareness of Industry 4.0 and circular practices by conducting a series of workshops. Further, the opportunities and risks are analyzed based on the business cases provided. The objective of this phase is to maximize the strengths and minimize the risks of the organization. The project scope is defined, and the project team is formulated. In general, the timeline for this phase is 1 to 3 months.

The second phase is Objective Setting, which is to evaluate key technological requirements and manage organizational considerations based on the project goals. In this phase, the existing competencies and the current business process are studied. For the identified improvement areas and requirements, a blueprint is developed. A benefit versus effort analysis is made and aligned with the stakeholders. In general, the timeline for the Objective Setting phase is 1 to 3 months.

The third phase is the Execution phase, where the real implementation of the Industry 4.0 and circular practices is conducted. In this phase, a detailed solution is developed, and a pilot study is performed. Based on the results, the solution is extended to other requirements. User acceptance testing is performed and confirmed. Later, training is organized with the workforce and the system goes live. Further, hyper-care support and sustainment are provided until the system is stabilized. It is advisable to start with a smaller unit and expand it to larger units so that the initial problems are identified and addressed at the early stage.

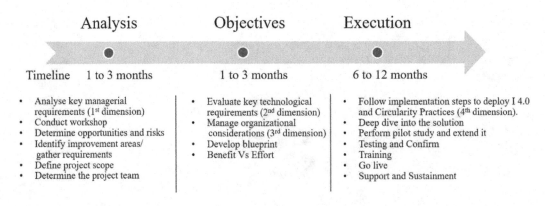

FIGURE 1.4   Structure of roadmap to Industry 4.0 and circularity transformation.

1.9 THEORETICAL AND PRACTICAL IMPLICATIONS

This chapter delivers multiple theoretical contributions to the existing knowledge on Industry 4.0 implementation. First, this is the first research of its kind to explore Industry 4.0 technologies that organizations can look for in a digital transformation journey. Second, a roadmap for digital transformation using Industry 4.0 tools is provided. In this context, this chapter aims to contribute to the Industry 4.0 literature.

The study offers practical implications for supply chain practitioners in the industrial scenario by leveraging Industry 4.0 technologies. A framework could facilitate industry practitioners to systematically prepare for Industry 4.0 challenges and develop competence to deal with global competition. The framework could provide insights to enable practitioners to systematically understand the concepts and technologies needed, the application domains, and the need for combining circular strategies with Industry 4.0. Appropriate training is essential for managers in understanding the significance of Industry 4.0 adoption in enhancing the organization's performance.

Stakeholders should pay attention to improvement areas so that they successfully adopt the proposed roadmap to implement Industry 4.0 technologies. Effective implementation of strategies makes the process more streamlined and moves towards greater flexibility, which in turn increases process and product development. Further, transforming into a digital organization improves the organization's reputation; therefore, there is a possibility to increase revenue with an increase in customers. Organizations find more visibility in their supply chain operations with Industry 4.0 technologies to make quick decisions, which results in improved stakeholder satisfaction.

1.10 CONCLUSION

Industry 4.0 redefines the integration of information and communication technology (ICT) with business processes. Supply chains are moving towards digital technologies to improve their supply chain operational efficiency. The impact of Industry 4.0 is realized by multiple functions of the supply chain, namely planning, procurement, manufacturing, logistics, and maintenance departments. The application of digital technologies such as the IIoT has improved the quality of the product by early detection of defective items; precise planning with integration analytics helps to predict manufacturing lead time, and automating supplier audits helps improve supplier performance.

In this chapter, a comprehensive body of literature is reviewed on digital supply chains and Industry 4.0 technologies. Based on the literature review and interaction with industry experts, the technologies that influence the organizational digital journey are studied. It is recommended that organizations have maturity in their supply chain process, and an appropriate IT infrastructure should also be in place before implementation of Industry 4.0 tools. Industry 4.0 technologies can be implemented in any type of organization, such as discrete, process-based, and project-based industries. A roadmap for digital transformation using Industry 4.0 tools is provided for organizations.

1.11 LIMITATIONS AND FUTURE RESEARCH DIRECTIONS

Implementing Industry 4.0 requires a specialized skill set and better IT infrastructure even for mature supply chain organizations. In the future, more case studies could be carried out for small and medium enterprises across varied sectors, thus expanding the practical validity of the framework. Researchers can expand the regions and study different sectors based on the strategic goals of the organizations.

REFERENCES

Abolghasemi, M., Beh, E., Tarr, G. and Gerlach, R. (2020), 'Demand forecasting in supply chain: The impact of demand volatility in the presence of promotion', *Computers & Industrial Engineering* **142**(1), 106380.

Ante, L. (2021), 'Digital twin technology for smart manufacturing and Industry 4.0: A bibliometric analysis of the intellectual structure of the research discourse', *Manufacturing Letters* **27**(1), 96–102.

Asamoah, D., Agyei-Owusu, B., Andoh-Baidoo, F. K. and Ayaburi, E. (2021), 'Inter-organizational systems use and supply chain performance: Mediating role of supply chain management capabilities', *International Journal of Information Management* **58**(1), 102195.

Birkel, H. S. and Müller, J. M. (2020), 'Potentials of Industry 4.0 for supply chain management within the triple bottom line of sustainability—A systematic literature review', *Journal of Cleaner Production* **289**(1), 125612.

Bressanelli, G., Perona, M. and Saccani, N. (2019), 'Challenges in supply chain redesign for the circular economy: A literature review and a multiple case study', *International Journal of Production Research* **57**(23), 7395–7422.

Corallo, A., Lazoi, M. and Lezzi, M. (2020), 'Cybersecurity in the context of Industry 4.0: A structured classification of critical assets and business impacts', *Computers in Industry* **114**(1), 103165.

de Vass, T., Shee, H. and Miah, S. (2021), 'IoT in supply chain management: Opportunities and challenges for businesses in early Industry 4.0 context', *Operations and Supply Chain Management: An International Journal* **14**(2), 148–161.

Erol, S., Jäger, A., Hold, P., Ott, K. and Sihn, W. (2016), 'Tangible Industry 4.0: A scenario-based approach to learning for the future of production', *Procedia CIRP* **54**(1), 13–18.

Fatorachian, H. and Kazemi, H. (2021), 'Impact of Industry 4.0 on supply chain performance', *Production Planning & Control* **32**(1), 63–81.

Florescu, A. and Barabas, S. A. (2020), 'Modeling and simulation of a flexible manufacturing system—A basic component of Industry 4.0', *Applied Sciences* **10**(22), 8300.

Garrido-Hidalgo, C., Olivares, T., Ramirez, F. J. and Roda-Sanchez, L. (2019), 'An end-to-end Internet of Things solution for reverse supply chain management in Industry 4.0', *Computers in Industry* **112**(1), 103127.

Gilchrist, A. (2016), *Introducing Industry 4.0*, Apress, Berkeley, CA.

Gupta, S., Modgil, S., Gunasekaran, A. and Bag, S. (2020, July), 'Dynamic capabilities and institutional theories for Industry 4.0 and digital supply chain', *Supply Chain Forum: An International Journal* **21**(3), 139–157.

Habib, M. S., Tayyab, M., Zahoor, S. and Sarkar, B. (2020), 'Management of animal fat-based biodiesel supply chain under the paradigm of sustainability', *Energy Conversion and Management* **225**(1), 113345.

Hahn, G. J. (2020), 'Industry 4.0: A supply chain innovation perspective', *International Journal of Production Research* **58**(5), 1425–1441.

Hernandez-de-Menendez, M., Morales-Menendez, R., Escobar, C. A. and McGovern, M. (2020), 'Competencies for Industry 4.0', *International Journal on Interactive Design and Manufacturing (IJIDeM)* **14**(4), 1511–1524.

Hofmann, E. and Rusch, M. (2017), 'Industry 4.0 and the current status as well as future prospects on logistics', *Computers in Industry* **89**, 23–34.

Huang, S. H., Liu, P., Mokasdar, A. and Hou, L. (2013), 'Additive manufacturing and its societal impact: A literature review', *The International Journal of Advanced Manufacturing Technology* **67**(5), 1191–1203.

Ivanov, D. and Dolgui, A. (2020), 'A digital supply chain twin for managing the disruption risks and resilience in the era of Industry 4.0', *Production Planning & Control* **32**(9), 775–788.

Jamil, F., Ahmad, S., Iqbal, N. and Kim, D. H. (2020), 'Towards a remote monitoring of patient vital signs based on IoT-based blockchain integrity management platforms in smart hospitals', *Sensors* **20**(8), 2195.

Kolberg, D. and Zühlke, D. (2015), 'Lean automation enabled by Industry 4.0 technologies', *IFAC-PapersOnLine* **48**(3), 1870–1875.

Lins, T. and Oliveira, R. A. R. (2020), 'Cyber-physical production systems retrofitting in context of Industry 4.0', *Computers & Industrial Engineering* **139**(1), 106193.

MacCarthy, B. L. and Ivanov, D. (2022), The digital supply chain—emergence, concepts, definitions, and technologies. In *The Digital Supply Chain* (pp. 3–24). Elsevier.

Manavalan, E. and Jayakrishna, K. (2019), 'A review of Internet of Things (IoT) embedded sustainable supply chain for Industry 4.0 requirements', *Computers & Industrial Engineering* **127**(1), 925–953.

Mastos, T. D., Nizamis, A., Terzi, S., Gkortzis, D., Papadopoulos, A., Tsagkalidis, N., . . . & Tzovaras, D. (2021), 'Introducing an application of an Industry 4.0 solution for circular supply chain management', *Journal of Cleaner Production* **300**(1), 126886.

Nunez-Merino, M., Maqueira-Marín, J. M., Moyano-Fuentes, J. and Martínez-Jurado, P. J. (2020), 'Information and digital technologies of Industry 4.0 and Lean supply chain management: A systematic literature review', *International Journal of Production Research* **58**(16), 5034–5061.

Oztemel, E. and Gursev, S. (2018), 'Literature review of Industry 4.0 and related technologies', *Journal of Intelligent Manufacturing* **31**(1), 127–182.

Pereira, M. M. and Frazzon, E. M. (2021), 'A data-driven approach to adaptive synchronization of demand and supply in omni-channel retail supply chains', *International Journal of Information Management* **57**(1), 102165.

Radziwon, A., Bilberg, A., Bogers, M. and Madsen, E. S. (2014), 'The smart factory: Exploring adaptive and flexible manufacturing solutions', *Procedia Engineering* **69**(1), 1184–1190.

RuBmann, M., Lorenz, M., Gerbert, P., Waldner, M., Justus, J., Engel, P. and Harnisch, M. (2015), 'Industry 4.0: The future of productivity and growth in manufacturing industries', *Boston Consulting Group* **9**.

Sahal, R., Breslin, J. G. and Ali, M. I. (2020), 'Big data and stream processing platforms for Industry 4.0 requirements mapping for a predictive maintenance use case', *Journal of Manufacturing Systems* **54**(1), 138–151.

Shamim, S., Cang, S., Yu, H. and Li, Y. (2017), 'Examining the feasibilities of Industry 4.0 for the hospitality sector with the lens of management practice', *Energies* **10**(4), 499.

Stock, T. and Seliger, G. (2016), 'Opportunities of sustainable manufacturing in Industry 4.0', *Procedia Cirp* **40**(1), 536–541.

Wang, G., Gunasekaran, A., Ngai, E. W. and Papadopoulos, T. (2016), 'Big data analytics in logistics and supply chain management: Certain investigations for research and applications', *International Journal of Production Economics* **176**(1), 98–110.

Wang, S., Zhang, C., Liu, C., Li, D. and Tang, H. (2017), 'Cloud-assisted interaction and negotiation of industrial robots for the smart factory', *Computers & Electrical Engineering* **63**(1), 66–78.

Wu, Y., Dai, H. N. and Wang, H. (2020), 'Convergence of blockchain and edge computing for secure and scalable IIoT critical infrastructures in Industry 4.0', *IEEE Internet of Things Journal* **8**(4), 2300–2317.

Yadav, G., Luthra, S., Jakhar, S. K., Mangla, S. K. and Rai, D. P. (2020), 'A framework to overcome sustainable supply chain challenges through solution measures of Industry 4.0 and circular economy: An automotive case', *Journal of Cleaner Production* **254**(1), 120112.

Yue, X., Cai, H., Yan, H., Zou, C. and Zhou, K. (2015), 'Cloud-assisted industrial cyber-physical systems: An insight', *Microprocessors and Microsystems* **39**(8), 1262–1270.

Zhang, Y., Qian, C., Lv, J. and Liu, Y. (2017), 'Agent and cyber-physical system based self-organizing and self-adaptive intelligent shopfloor', *IEEE Transactions on Industrial Informatics* **13**(2), 737–747.

2 The Autonomy of Autonomous Robots

Raunika Anand, Arijeet Nath, and S. Aravind Raj

CONTENTS

2.1 Introduction 9
   2.1.1 Evolution of Robots 9
2.2 Classification of Autonomous Robots 10
2.3 Software Algorithms in Autonomous Robots 10
   2.3.1 Mapping in Robotic Navigation 11
   2.3.2 Path Planning in Robotic Navigation 13
2.4 Application of Autonomous Robots 14
2.5 Autonomous Robots in Industry 15
2.6 Collaborative Robots in Industry 16
2.7 Conclusion 16
References 17

2.1 INTRODUCTION

The word "robot" was coined by Karel Capek in 1920 in his science-fiction play, R.U.R. (Rossum's Universal Robots). Derived from the Czech "robota," the word meant servitude [1]. The concept of robot was introduced by him to convey his protest against the rise of modern technology by showing the growth of robots and their eventual revolt against humans. While the concept of robots has been around for a long time, it was during the 1940s that the first modern-day autonomous robot were born [2]. It was during 1948–1949, the first electronic autonomous robots with complex behavior were developed by William Grey Walter. These robots, named Elmer and Elsie, through phototaxis, autonomously went to a recharging station [3, 4].

Robots were a concept of science fiction up until the 1950s. Since then, robots have been used in various applications from industries to deep sea navigation and from navigating the earth to extraterrestrial exploration. All of these use cases of robots were aimed towards adding to human functions and performing tasks that were deleterious to humans. There is no one type of robot, and various classifications are given to define these machines.

2.1.1 EVOLUTION OF ROBOTS

Robots have come a long way. Robots today are controlled by computers that are powerful enough to simulate the mind of an insect, although they are expensive and therefore find very few profitable niches in society. But these applications are galvanizing research and building a base for huge future growth. Robot evolution in the direction of full intelligence will accelerate. There are four categories for robots: first generation, 1950–1967; second generation, 1968–1977; third generation, 1978–1999; and fourth generation, 2000–present.

The first-generation robots are the ones that are programmable machines, in which the real modes of execution cannot be controlled, and they have zero communication with the outside environment. These robots use rudimentary equipment and do not have servo-controllers. Most of these robots

DOI: 10.1201/9781003186670-2

are pneumatic, and the automatic regulators consist of logic gates that can be actuated through air. The first generation robots were primarily used for loading and unloading purposes or for carrying out handling operations and pick-and-place operations.

The second-generation robots are programmable machines that can self-adapt and have basic cognizance of the external environment. These robots have servo-controllers and can be programmed so they move from either point to point or on a continuous path. They are controlled by programmable logic controllers and minicomputers. They can perform much more complex tasks compared to the first-generation robots.

The third-generation industrial robots consist of self-programmable machines that can interact with the outside environment and operator using voice commands, vision, and so on. They possess some ability to be reprogrammable for the execution of a given task. These machines operate under servo control and again can be programmed to move either point to point or on a continuous path. The third-generation robots can also perform sort of "intelligent" tasks. They can also perform other complex tasks such as tactile operations and freehand machining.

In the fourth-generation industrial robots, the abilities of the robots reach an even higher level where they can perform advanced computing, logical reasoning and confounding control strategies, and collaborative behavior with humans and other robots.

2.2   CLASSIFICATION OF AUTONOMOUS ROBOTS

One of the most distinct classification of robots is a fixed robot versus a mobile robot. Since the workspaces of these two types of robots are very different, their functions and capabilities are different [5]. A fixed robot works in a well-defined and confined workspace to perform repetitive tasks efficiently. These types of robots are automated industrial robots such as Unimate. A mobile robot is a type of robot that moves around its environment and carries out tasks in the environment, avoiding obstacles. A perfect example of this kind of robot would be a Roomba robot, which is used for vacuuming the floor. This robot needs to deal with a far more complex and unpredictable environment compared to a fixed robot and thus needs to be far more autonomous.

Robots are also classified as autonomous robots or telerobots based on the level of autonomy [6]. A major difference between autonomous and tele-operated robots is the level of autonomy. Tele-operated robots are the least autonomous and always require an operator while performing a set of tasks. Traditional industrial robots are a type of tele-operated robot because they can be seen doing repetitive tasks such as welding, cutting, transfer of parts, and so on. A robot with high autonomy can make decisions and perform real-world tasks with minimal human intervention.

For traditional industrial robots to function efficiently and prevent accidental injuries to humans, they must be separated from human contact and confined to human-free cages. With improvements in physical human–machine interaction, such robots are also designed to work around humans. Such robots are known as collaborative robots or cobots. Using different types of sensors such as tactile, UV, and infrared and robust control algorithms, a cobot detects humans and avoids obstacles using AI [7]. Cobots are robots with a high level of autonomy.

2.3   SOFTWARE ALGORITHMS IN AUTONOMOUS ROBOTS

Various machine learning algorithms have been used in the control of autonomous robots. Using these algorithms enables the control and navigation of the robot using the software systems [8]. Among the most prominent parts of autonomous robot navigation is the path planning system. It points to path determination that is free of collision from start to target point based on evaluation criteria so that the total cost of the corresponding path is reduced [9]. For successful execution of navigation, the robot must carry out three primary tasks: mapping of the environment, path planning, and driving [10]. The following context will focus on the different algorithms used for mapping and path planning in autonomous robot navigation.

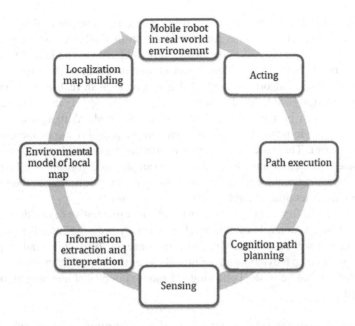

FIGURE 2.1   Steps in mobile robot navigation [12].

Autonomous robot navigation is categorized into three types, global, local and personal navigation [11]. Global navigation requires prior knowledge of the environment and thus depends on the knowledge by which the robot can determine the position of different objects in the environment. Local navigation requires the robot to calculate the position of dynamic objects in its vicinity and change its position and orientation (pose) based on the inputs received from different types of sensors such as ultrasonic sensors; infrared sensors; laser imaging, detection, and ranging (LIDAR); and so on. Handling of different objects in the environment with respect to each other based on their pose is called personal navigation.

2.3.1   Mapping in Robotic Navigation

Mapping of the environment is one of the three criteria that must be fulfilled in order to have successful navigation. Depending on the algorithms to be used for mapping, different factors are considered. For example, one algorithm can be incremental, meaning that it can be run in online or real-time mode, while others require that the data of the environment be processed multiple times. Some algorithms require an exact pose of elements in the environment fed to it, while others can create a virtual map of the environment based on sensor data. One thing these algorithms have in common is that they are all probabilistic [13]. The reason is that mapping is characterized by uncertainty and noise from sensors. To accommodate the noise and uncertainty, researchers use probabilistic algorithms, which are capable of modeling the different sources of noise and calculate the effects they have on mapping and localization. Environment mapping and localization of the robot are both imperative in mapping the environment as well as pinpointing where the robot actually is in the environment. But both processes have uncertainties in them, and focusing on the one introduces noise from the other. As a result, it is imperative to map an unknown environment while at the same time localizing the robot in the unknown environment. This problem is called simultaneous localization and mapping (SLAM) [14].

There are three types of mapping: probabilistic, incremental, and hybrid methods. The main aim of the probabilistic type of algorithm is to solve the correspondence problem. This is where the robot should ascertain if data taken at different times are of the same physical object. Probabilistic techniques produce very accurate results compared to other methods. A major drawback, however, is that most of them are quite intensive computationally to be able to run in real time. Incremental methods are developed to function in a much easier fashion in real time than probabilistic methods. They are a more reasonable algorithm method that can be used in real-time autonomous robots. Hybrid algorithms are designed to handle much more complex mapping in real time. These consist of a combination between probabilistic and incremental methods. Although newer and difficult to implement, they provide much better yields than simply probabilistic or incremental methods for an autonomous robot. There are certain features that can be desirable for the problem they are attempting to solve or the environment to map. For example, an algorithm that helps create maps incrementally and does not require several passes through data is desirable. In the same way, some algorithms require an exact pose of a robot, while others do not.

To select an algorithm, first consider the constraints. It is imperative to decide if real-time mapping is required or not. Second, know the level of accuracy required: whether the maps created based on the data input will be stationary or for a simultaneous localization and mapping (SLAM) system. It is optional to map a dynamic environment. After the primary constraints and algorithms are defined, a selection can be made. Following are some examples of mapping algorithms used in robotic navigation.

1. Bayes' filter: The fundamental of any probabilistic algorithm is Bayes' rule. Bayes' rule says that the probability of x given data, d (the posterior) is equal to the probability of d with respect to x (likelihood) multiplied by probability of the previous x, divided by the probability of d, which is the evidence or the marginal likelihood shown in Equation 1. In other words, this means that the present belief can be changed or is dependent on the previous observations. Bayes' filter uses a continuous estimation to approximate the map as well as the robot's pose

$$P\frac{x}{d} = \frac{P\left(\dfrac{d}{x}\right) \times P(x)}{P(d)} \qquad\qquad \text{Equation (1)}$$

2. Kalman filter–based algorithm: Kalman filter–based algorithms are often known as SLAM algorithms. In SLAM, the requirement of the work environment doubled [15]. This means that the map is often used for supporting other factors. It is formed to restrict any measurement errors that occur during the state of the robot. Second, it helps the robot to reduce localization errors using a set of key points in the map. As a result, SLAM is generally preferred in areas where prior knowledge of the environment is not necessary. The Kalman filter–based algorithm has three assumptions. First, the following state function (motion model) should be a linear function with added Gaussian noise. A linear function simply means the robot pose and map at a given time should correspond linearly to the previous robot pose and map. This is true for the map; however, the relation between adjacent robot pose states is nonlinear. As a result, to make them linear, the Kalman filter approximates the motion model. This Kalman filter is thus known as the extended Kalman filter [16]. Gaussian noise is nothing but an approximation of nonlinear sensor noise. Second, the same characteristics must also apply to the perceptual model. And third, the initial uncertainty must be Gaussian. This can be a problem since errors such as incorrect tuning in noise during approximation of as Gaussian can result in divergence in the filter, thus destabilizing the overall system [17]. The advantage of using a Kalman filter is it is ideal for dynamic environments. They are computationally less intensive and fast, as they do not store memory, unlike the previous states and thus are well suited for real-time operations.

3. Monte Carlo localization (MCL): Developed by Frank Dellaert in 1999, this method aimed to overcome the drawbacks of the Kalman filter approach. As mentioned, probabilistic approaches to localization are the most promising, and the Kalman filter algorithms have been proven robust and accurate. But they have some problems; for example, they cannot represent ambiguity and cannot relocalize the robot in case of initial failures. In the traditional probabilistic approach, uncertainty is described using the probability density function. However, in Monte Carlo localization, instead of doing that, the probability density is described by picking sample sets randomly. Using this type of sample-based representation, a localization method that can represent random distributions is obtained [18]. With this method, the algorithm can accommodate arbitrary noise distributions and nonlinearities, which is difficult with the Kalman filter due to the approximation of sensor noise as Gaussian noise. Thus, Monte Carlo localization avoids the requirement to extract features from sensor data. Some of the advantages of using MCL are as follows [19]:

   a. As opposed to Kalman filtering techniques, it represents multi-modal distributions and hence is capable of globally localizing a robot.
   b. It is more precise than Markov localization, having a fixed cell size, as the state represented in the samples is not represented in discrete quantities.
   c. It is simple and easy to use.

2.3.2 Path Planning in Robotic Navigation

Selecting the appropriate algorithm for each stage is imperative for the smooth functioning of the navigation system. A general path planning algorithm must satisfy the following criteria [20]:

1. The corresponding path must cost the least to prevent indirection.
2. It should be quick as well as accurate so that it does not frustrate the simulation process. It must be robust in case of zero collisions during the navigation process.
3. It should be generalized so it can be used in different maps and not be tailored to one specific map.

Some path planning algorithms are described in the following:

1. A* algorithm: It is one of the most well-known path planning algorithms, which can be applied on topological or metric C-space or configuration space [21]. Metric and topological are two mapping methods; here metric mapping means to describe the geometric properties of the mapped environment, whereas the topological method describes the connectivity of several places [13]. Developed by Peter Hart, Bertram Raphael, and Nils Nilsson of Stanford Research Institute in 1968 [22], this algorithm combines the shortest path and heuristic searching. It is defined as a best-first algorithm, as every cell in the C-space is evaluated by a value, as shown in Equation 2.

$$f(v) = h(v) + g(v) \qquad \text{Equation (2)}$$

Here $h(v)$ is the heuristic distance of a cell up to the target state, and $g(v)$ is given by the path length from initial to the target state through a given cell sequence. The value of every adjacent cell of the cell that is reached is calculated by $f(v)$. The cell with the lowest $f(v)$ is chosen as the next cell in the sequence [23]. A major advantage of using the A* algorithm is that the algorithm can quickly converge, while a major disadvantage is that it is computationally intensive.

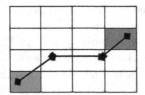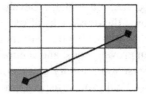

FIGURE 2.2   The A* algorithm at work [23].

B. Rapidly exploring random tree (RRT): This algorithm was developed by Steven M. LaValle and James J. Kuffner Jr [24]. RRT is a type of sampling-based planning (SBP) algorithm. SBP algorithms have been widely used for mobile robot path planning. SBP algorithms can provide quick solutions for difficult and high-dimensional problems using random sampling in the search space [25]. An RRT is designed with parameters that are as random as possible. This leads to better performance analysis and consistent behavior. One major advantage of RRTs is that they can be directly applied to nonholonomic planning. This arises from the fact that RRTs require no connections to be made between configuration pairs, while probabilistic roadmaps typically require tens of thousands of connections to be made between them [26].

C. Genetic algorithm: This algorithm was originally developed in 1975 by Bremermann [27], who was the pioneer of evolutionary computation. The algorithm itself is an evolutionary computational approach used to optimize a given function for maximizing and minimizing operation. It is the semblance of the process of finding out the fittest solution through reproduction and natural selection. In this process of evolution, the randomness present in the genetic algorithm (GA) is to set to one to control the randomization. The genetic algorithm is more efficient and robust than other random search algorithms without needing extra information about the given problem. This feature of the GA helps to calculate solutions easily, which other optimization algorithms cannot do due to deficiency of continuity and linearity. The ability to get a global optimum and high parallelism make it efficient.

2.4   APPLICATION OF AUTONOMOUS ROBOTS

Autonomous robots and systems are not a new concept. They have been an integral part of modern technology in controlling railroads, autopilot systems on airplanes, and maintenance sectors in railways and marine vehicles, to name a few. Autopilot technology uses gyroscopic technology to control and maintain speed and altitude. Auto-tillers in sailboats are used to maintain a given course based on feedback from wind. Automation in industrial sectors is also cost effective in the long run. Over the period from 1990 to 2005, the mean quality adjusted price of industrial robots reduced by 80 percent in the European Union, United States, and United Kingdom [28]. Monotonous, manual tasks in traditional industries carried out by humans are likely to cause errors in, for example, assembly, inspection, maintenance, and so on. These errors can be detrimental to the industry and can cause unforeseen loss in productivity, revenues, and even lives. Autonomous systems, on the other hand, are known to have high accuracy and efficiency in the tasks they perform. They can routinely perform monotonous tasks with the same accuracy for extended periods of time, making them more efficient and cost effective in the process.

The role of autonomous rovers has also been essential in realizing the mission of space exploration. The Pathfinder mission of the National Aeronautics and Space Administration (NASA) paved the way for future planetary exploration with rovers. The Prop-M rover in the Mars 2 and 3 missions in 1971 was the first ever rover to land on Mars. It was connected to the lander with a 15-meter umbilical cord. It was included for the mission to measure the soil density, for which it carried a dynamic penetrometer and radiation densitometer. The rover, designed to move on skis, was inevitably autonomous because the delay in sending radio signals from Earth to Mars made it difficult to control the rover manually. It used two elementary impact bars on the front to detect rocks and other obstacles. Whenever these bars came in contact with obstacles, they signaled the rover to retreat and turn. It stopped and took measurements every 1.5 meters. The missions, however, failed and Prop-M could never actually make any measurements on Mars, but the missions were a vital step towards automation. The Sojourner rover launched in 1997 during the Pathfinder mission was the first wheeled rover to ever land on Mars and was the foundation that paved the way to Mars rovers today. Known as the Microrover Flight Experiment (MFEX), the mission aimed to be a flight experiment for autonomous technologies, whose main mission was to determine rover performance in the crudely understood terrain of Mars. The autonomous navigation and obstacle avoidance used for the rover was rather elementary. A "Go to Waypoint" command was implemented for autonomous navigation of the robot. This command was given by a member of the operations team known as the robot driver. Using the camera images from the rover and the rover images from the lander camera, a rough terrain of the nearby areas was identified by the robot driver, by which he/she identified the x and y coordinates of the rover location and the destination coordinates. Once these values were fed to the rover, it started to move in an approximate straight line towards the destination. Being primarily a technology experiment itself, the mission was an astounding success and rightly named, as it paved the way for future Martian missions. With the advancement of digital control of electronics during the 1970s, advancements were made in autonomous systems due to the increased interest in cognition and perception in the emerging field of artificial intelligence. These autonomous robots were able to perform complex tasks with little or no human interaction. With the cost of hardware components like sensors and processors decreasing, the growth of autonomous systems has been significant in all walks of life.

During the unprecedented times of COVID-19, autonomous systems have seen a boost in healthcare. The robot known as the smart tissue autonomous robot (STAR) performed a delicate procedure called intestinal anastomosis in which intestines that have been cut are sewn back together such that the sutures are evenly spaced and taut to prevent leakage. Its consistency and accuracy surpassed those of two surgeons who were assigned with the same task. Using a near-infrared fluorescent (NIRF) light, it tracks small marks placed on the tissue beforehand to visualize the path, and it is supposed to cut and adjust the cutting tool movement accordingly. The da Vinci surgical system is one of the prime examples of robotic surgery systems that have been implemented in hospitals across the world, but it cannot operate independently. Robots such as STAR are much more autonomous and can independently perform tedious tasks such as suturing with precision and consistency. Robots developed by UVD Robots have seen large-scale growth in usage during the current COVID-19 pandemic. Using ultraviolet-C rays to disinfect hospital floors, these robots hinder the spread of germs and viruses in hospitals. Model C, which is the third generation of ultraviolet disinfection (UVD) robots, is also autonomous. It uses LIDAR sensors to scan its surrounding environment and creates a digital map of the environment. The operator then pinpoints positions in the map to carry out the disinfection process. The robot then navigates around the room independently using a computational problem called simultaneous localization and mapping, which essentially allows the robot to update the map of the environment and also keep track of its own position at the same time. It can also automatically pause the disinfection process if it detects a human in the room, as UV rays can be harmful to humans.

2.5 AUTONOMOUS ROBOTS IN INDUSTRY

Industry 4.0 is the latest stage of the industrial revolution that has revolutionized the way industries operate. With the help of advanced digitalization in industries and IOT, the means of production and manufacturing have noticed a new shift in paradigm where industries are changing from mass production to customized production systems [29]. The fourth industrial revolution aims to integrate the virtual world with the physical world where everything from smart machines, work-ers, and tools to customers, products, and services is interconnected [30]. With humans at the core of this new shift in paradigm, it becomes imperative to take into consideration human factors to successfully implement it. Hence implementing such systems and robots in the workplace environ-ment needs to consider and incorporate the human elements to maintain the safety of humans as well as robots [31].

Introducing robots in industrial settings where they have to work in close proximity to humans poses two major problems: First, the robots should be designed in such a way that they cause no harm to the human workers nearby. Second, problems that factor in the human mindset also need to be considered. For example, workers who typically are not expert in handling complex robotic systems are underconfident in operating them, which gives rise to fear and anxiety [32]. To address such issues, many industrial robotics manufacturing companies have taken steps to develop bet-ter and more efficient collaborative robots. Collaborative robots are one of the primary drivers of Industry 4.0. These robots are much more efficient, flexible, and most importantly safe.

2.6 COLLABORATIVE ROBOTS IN INDUSTRY

A collaborative robot functions such that it can share a common workspace with a human. With prime focus given to the safety of its human operator, other factors like ease in programming, ease in deployment, and being lightweight are taken into consideration for a robot to be considered collaborative. Such robots are necessary in the workspace environment, as it has the best of both worlds: physical, repetitive tasks can be done by robots continuously for long hours, while using the cognition of human minds brings versatility and innovation to completing a task [33]. The primary aim for manufacturing in the future is greater flexibility and custom batch sizes of pro-duction. This was particularly difficult in traditional industries, which were rigid and difficult to configure. Such industries were slowly supplanted by the development of robots such as the KUKA (Leichtbauroboter—intelligent industrial work assistant) LBR iiwa. It is a lightweight industrial robot with seven axes. All of its joints have torque and position sensors with which it controls the impedance and position. This enables the robot to move quickly with respect to the process forces and interact safely with humans [34].

Installing autonomous robots in industries will make it easier to transport goods across the dif-ferent sectors of industry with more flexibility and reduced cost. This will then enable increased production, promote industrial growth, foster human safety, boost the economy, and ultimately change the dynamics of the industrial system. The future industry holds the possibility of more flexible and cooperative autonomous robots that can interact not only with one another but also with humans and work side by side along with them [35].

2.7 CONCLUSION

Autonomous systems have the capability to scale up human abilities and go further. Such systems can work in deleterious places such as construction sites, coal mining areas, the deep sea, and even deep space. There have been apprehensions regarding autonomous robots supplanting human work-ers and thereby taking away jobs, but instead, if anything, they have made our lives easier and more efficient. With the aid of robots such as ultraviolet disinfection robots, hospital staff can focus on other important tasks. Surgical robots can perform tedious and repetitive tasks for long hours with consistency and without fatigue, thereby making the job of surgeons easier. Installing such systems in hospitals for surgery also eliminates the margin for human error. Unmanned aerial vehicles and swarms of drones have been shown to have a 99 percent sensor coverage rate under two hours over 2 square kilometers of area simulated after real tsunami locations, which is well under the first 72-hour window of a search and rescue operation, which is deemed critical. These examples show the importance of autonomous systems in human lives. Also, the growth in industries based on the fourth-generation industrial revolution called Industry 4.0 necessitates a connected and remote working environment, which can done in a better way using an autonomous robots. Research is ongoing in the field of combining cyber-physical systems (CPSs) and robotics, which results in exemplary outcomes in all industrial sectors. Autonomous systems have impacted every walk of life, and their future implications are going to be even more massive.

REFERENCES

1. Reilly, K. (2011). From automata to automation: The birth of the robot in RUR (Rossum's Universal Robots). In *Automata and Mimesis on the Stage of Theatre History* (pp. 148–176). Palgrave Macmillan, London.
2. Hagis, Christopher. (2003). *History of Robots* (Doctoral dissertation, Master's thesis, Wagner College).
3. Walter, W. G. (1950). An imitation of life. *Scientific American, 182*(5), 42–45.
4. Holland, O. (2003). Exploration and high adventure: The legacy of Grey Walter. *Philosophical Transactions of the Royal Society of London. Series A: Mathematical, Physical and Engineering Sciences, 361*(1811), 2085–2121.
5. Ben-Ari, M., & Mondada, F. (2018). Robots and their applications. In *Elements of Robotics* (pp. 1–20). Springer, Cham.
6. Choi, J. J., Kim, Y., & Kwak, S. S. (2014, August). The autonomy levels and the human intervention levels of robots: The impact of robot types in human-robot interaction. In *The 23rd IEEE International Symposium on Robot and Human Interactive Communication* (pp. 1069–1074). IEEE.
7. She, Y., Song, S., Su, H. J., & Wang, J. (2020). A comparative study on the effect of mechanical compliance for a safe physical human–robot interaction. *Journal of Mechanical Design, 142*(6).
8. Mainampati, M., & Chandrasekaran, B. (2020, January). Evolution of machine learning algorithms on autonomous robots. In *2020 10th Annual Computing and Communication Workshop and Conference (CCWC)* (pp. 0737–0741). IEEE.
9. Li, L., & Liu, L. (2009). An algorithm of multi-robots local collision avoidance. *Ordnance Industry Automation, 6*(6), 62–65.
10. Sariff, N., & Buniyamin, N. (2006, June). An overview of autonomous mobile robot path planning algorithms. In *2006 4th Student Conference on Research and Development* (pp. 183–188). IEEE.
11. Patle, B. K., Pandey, A., Parhi, D. R. K., & Jagadeesh, A. (2019). A review: On path planning strategies for navigation of mobile robot. *Defence Technology, 15*(4), 582–606.
12. Patle, B. K. (2016). *Intelligent Navigational Strategies for Multiple Wheeled Mobile Robots Using Artificial Hybrid Methodologies* (Doctoral dissertation).
13. Thrun, S. (2002). *Robotic Mapping: A Survey.* School of Computer Science, Carnegie Mellon University, Pittsburgh, PA.
14. Saeedi, S., Trentini, M., Seto, M., & Li, H. (2016). Multiple-robot simultaneous localization and mapping: A review. *Journal of Field Robotics, 33*(1), 3–46.
15. Raja, M. (2019). Application of cognitive radio and interference cancellation in the L-band based on future air-to-ground communication systems. *Digital Communications and Networks, 5*(2), 111–120.
16. Julier, S. J., & Uhlmann, J. K. (1997, July). New extension of the Kalman filter to nonlinear systems. In *Signal Processing, Sensor Fusion, and Target Recognition VI* (Vol. 3068, pp. 182–193). International Society for Optics and Photonics, Bellingham, WA.
17. Santhanakrishnan, M. N., Rayappan, J. B. B., & Kannan, R. (2017). Implementation of extended Kalman filter-based simultaneous localization and mapping: A point feature approach. *Sādhanā, 42*(9), 1495–1504.
18. Thrun, S., Fox, D., Burgard, W., & Dellaert, F. (2001). Robust Monte Carlo localization for mobile robots. *Artificial Intelligence, 128*(1–2), 99–141.
19. Dellaert, F., Fox, D., Burgard, W., & Thrun, S. (1999, May). Monte Carlo localization for mobile robots. In *Proceedings 1999 IEEE International Conference on Robotics and Automation (Cat. No. 99CH36288C)* (Vol. 2, pp. 1322–1328). IEEE.
20. Niederberger, C., Radovic, D., & Gross, M. (2004, June). Generic path planning for real-time applications. In *Proceedings Computer Graphics International, 2004* (pp. 299–306). IEEE.
21. Cui, S. G., Wang, H., & Yang, L. (2012, October). A simulation study of A-star algorithm for robot path planning. In *16th International Conference on Mechatronics Technology* (pp. 506–510).

22. Hart, P. E., Nilsson, N. J., & Raphael, B. (1968). A formal basis for the heuristic determination of minimum cost paths. *IEEE Transactions on Systems Science and Cybernetics*, *4*(2), 100–107.
23. Duchoň, F., Babinec, A., Kajan, M., Beňo, P., Florek, M., Fico, T., & Jurišica, L. (2014). Path planning with modified A star algorithm for a mobile robot. *Procedia Engineering*, *96*, 59–69.
24. LaValle, S. M., & Kuffner Jr, J. J. (2001). Randomized kinodynamic planning. *The International Journal of Robotics Research*, *20*(5), 378–400.
25. Elbanhawi, M., & Simic, M. (2014). Sampling-based robot motion planning: A review. *IEEE Access*, *2*, 56–77.
26. LaValle, S. M. (1998). *Rapidly-Exploring Random Trees: A New Tool for Path Planning*.
27. Bremermann, H. J. (1958). *The Evolution of Intelligence: The Nervous System as a Model of Its Environment*. Department of Mathematics, University of Washington, Seattle, WA.
28. Vithanage, R. K., Harrison, C. S., & DeSilva, A. K. (2019). Importance and applications of robotic and autonomous systems (RAS) in the railway maintenance sector: A review. *Computers*, *8*(3), 56.
29. Lasi, H., Fettke, P., Kemper, H. G., Feld, T., & Hoffmann, M. (2014). Industry 4.0. *Business & Information Systems Engineering*, *6*(4), 239–242.
30. Bragança, S., Costa, E., Castellucci, I., & Arezes, P. M. (2019). A brief overview of the use of collaborative robots in Industry 4.0: Human role and safety. *Occupational and Environmental Safety and Health*, 641–650.
31. Maurice, P., Padois, V., Measson, Y., & Bidaud, P. (2017). Human-oriented design of collaborative robots. *International Journal of Industrial Ergonomics*, *57*, 88–102.
32. Ferraguti, F., Pertosa, A., Secchi, C., Fantuzzi, C., & Bonfè, M. (2019, March). A methodology for comparative analysis of collaborative robots for Industry 4.0. In *2019 Design, Automation & Test in Europe Conference & Exhibition (DATE)* (pp. 1070–1075). IEEE.
33. Landi, C. T., Villani, V., Ferraguti, F., Sabattini, L., Secchi, C., & Fantuzzi, C. (2018). Relieving operators' workload: Towards affective robotics in industrial scenarios. *Mechatronics*, *54*, 144–154.
34. Mokaram, S., Aitken, J. M., Martinez-Hernandez, U., Eimontaite, I., Cameron, D., Rolph, J., . . . & Law, J. (2017). A ROS-integrated API for the KUKA LBR iiwa collaborative robot. *IFAC-PapersOnLine*, *50*(1), 15859–15864.
35. Rüßmann, M., Lorenz, M., Gerbert, P., Waldner, M., Justus, J., Engel, P., & Harnisch, M. (2015). Industry 4.0: The future of productivity and growth in manufacturing industries. *Boston Consulting Group*, *9*(1), 54–89.

3 Smart Technologies for Industry 4.0 and Its Future

P. Ben Rajesh and A. John Rajan

CONTENTS

3.1 Introduction 20
3.2 Base Work of This Book Chapter 21
3.3 Big Data 21
3.4 Characteristics of Big Data 21
    3.4.1 Velocity of Big Data 21
    3.4.2 Volume of Big Data 21
    3.4.3 Veracity of Big Data 21
    3.4.4 Variety of Data 22
    3.4.5 Value of Data 22
    3.4.6 Variability of Data 22
    3.4.7 Scalability of Data 22
    3.4.8 Relational Data 22
3.5 Applications of Big Data in Logistics 22
    3.5.1 Effective Planning of Cargo Circuits 22
    3.5.2 Warehouse Automation 23
    3.5.3 Monitoring Product Quality 23
    3.5.4 Cargo Mobility Report 23
    3.5.5 Electrifying Last-Mile Delivery 23
3.6 Artificial Intelligence 23
    3.6.1 Purpose of AI 24
    3.6.2 AI Manages Redundant Data in a Manufacturing and Supply Chain—
        Case Study 24
    3.6.3 Significance of AI in Production and Supply Chain Visibility—
        Case Study 24
3.7 Machine-to-Machine Technology 25
    3.7.1 Process of M2M 25
    3.7.2 Applications of M2M 25
    3.7.3 Features of M2M 26
    3.7.4 M2M in Smart Logistics 26
3.8 Digitization 26
    3.8.1 Benefits of Digitization 27
    3.8.2 Significance of Digitization 27
    3.8.3 Digitized Logistics Evolution—A Study 27
    3.8.4 Applications of Logistics 4.0 27
3.9 Conclusion 28
References 28

DOI: 10.1201/9781003186670-3

3.1 INTRODUCTION

Digitization is the buzzword of the 21st century and became an industrial and supply chain trend. This trend generates more job creation, attracts higher investments, and improves the living standards of consumers. Moreover, digitization extends customization to the reach of customers (Buyukozkan & Gocer, 2018). Popular shoe manufacturers (Adidas, Nike) have offered a customer-centric design interface, where customers have freedom to design their shoes. These digital platforms allow end users to purchase the finished product at any time or location. Digitization enables consumers with platforms, like e-commerce, social media, and mobile networks, to evaluate and shop as they desire. In addition to that, digitization reinforced marketing departments by gathering data on consumer purchasing behavior, purchasing time frequency, items they purchase, and purchase quantity (Buyukozkan & Gocer, 2018)

Digitization (Geisberger & Broy, 2012) enables the dream word "Industry 4.0" to be a real process. Digital natives born after the millennium have higher expectations than Baby Boomers. They need highly reliable products with affordable prices, made with eco-sensible raw materials with low lead time and post-delivery support (Kearney, 2015), too. With multiple constraints, manufacturers would launch their product in the market space only through strategic planning and investment in automation technology. The customer's instant happiness was the primary objective of ultra-age corporates. The four main pillars of Industry 4.0 are the Internet of Things (IoT), Industrial Internet of Things (IIoT), cloud-based manufacturing (Attaran, 2017), and smart manufacturing, which have transformed manufacturing to be fully digitized and intelligent (Chase, 2019).

Industry 4.0 (Vaidya et al., 2018) offers the following privileges: installation phase, migrating cost (i.e., cost of adapting Industry 4.0 into existing infrastructure), enhancing profitability and productivity (Fruhlinger, 2020), transparency in process (more visibility with layers of operations), flexibility, calculation speed, simplifying complex tasks, improvement of customer satisfaction and customer experience, real-time data monitoring and tracking, and self-learning machines (Fruhlinger, 2020).

Industry 4.0's primary objective was to develop visibility among the layers of manufacturing with data on machine running hours, down time, and production efficiency. It was also used in developing flexible manufacturing such as mass production and job manufacturing using 3D printing machines. Moreover, Industry 4.0 was used for predicting manufacturing lead time with 95% accuracy. It also helped to categorize entire tasks into multiple layers of simplified tasks with lower degrees of complexity. It also helped the workforce detect higher-order complex tasks that demand more attention. The key motivation for adopting Industry 4.0 was quicker implementation (in the range of a week) with less modification of existing infrastructure (i.e., <20%). The cost of implementation was highly affordable when compared to capital investment in conventional devices.

This chapter examines the smart technologies of Industry 4.0 used in the logistics domain. It records the latest case studies from the leading logistic movers (DHL, Alibaba, UPS, FedEx), who considered Industry 4.0 a good strategy to balance the demand when national and international borders were closed during the pandemic. A methodological approach is used to gather data through digital surveys and web articles on leading logistic players across the world. The digital survey provides insights on new corporate strategies to enhance their investment in digital infrastructure to manage the new normal, such as social distancing, travel restrictions, quarantine duration, and on. Drones, self-driving cars, tracking shipments with Bluetooth, barcode systems, 3D printing, voice bots, logistic control towers (to solve international logistics issues), and smart apps enabled logistic domains to continue to work during the pandemic.

In a statistical survey, "Public Perception on Robotics and AI Applications in Europe," 37% of consumers preferred robotics and AI as an efficient way to transport goods, and 34% people considered robotics good for society. Though there are a few challenges (Evas, 2017), such as threats to privacy, to humanity, and to fundamental rights, the merits surpass the disadvantages. Digital technology enablers of Industry 4.0 are big data, machine-to-machine (M2M) interfaces, advanced analytics with the aid of machine learning algorithms, and the Internet of Things. This chapter's objective is to investigate the research query: Is Industry 4.0 a convenient tool in the logistics and supply chain?

3.2 BASE WORK OF THIS BOOK CHAPTER

The authors of this chapter conducted a review of literature, indexed journals, digital magazines, online logistic societies (DC Velocity, Supply Chain Brain), and industrial reports with keywords "Industry 4.0," "Digital enablers," "Big data in Logistics," and "AI in logistics." These articles were selected to determine essential information and answers for the research query. Our literature review converged towards "Industry 4.0" in logistics (Kuckelhaus & Chung, 2018). Articles and journals were preferred from supply chain and logistics journals. Articles were downloaded from the following digital databases: Google Scholar, Research Gate, Academia, and so on, and the search was refined with year of publication (not earlier than 2016). The review outcome is a detailed summary of real-world applications of big data, AI, and M2M with business world case studies.

3.3 BIG DATA

The term "big data" refers to high-volume data that is too extensive, rapid, and highly complicated to execute using conventional methodology. Therefore, the process of retrieving and saving an enormous quantity of data for analytics has been a challenge for data professionals. Moreover, big data comprises (Jeske et al., 2013) different categories of data: *unstructured, semi-structured*, and *structured data*. However, professionals aim mainly to use unstructured data because of its complex sorting nature. The size of big data is dynamic, and its storage size ranges from a few terabytes to zettabytes of data.

Some examples of big data are the data captured in stock exchange institutions, which could generate at least 1 terabyte of data trade per day, and video uploads, messages, and photo sharing through social media apps (such as Facebook and WhatsApp) produce more than 500 terabytes of data every day (Digital Report, 2020).

3.4 CHARACTERISTICS OF BIG DATA

Big data characteristics are defined by the variables velocity, volume, veracity, variety, value, variability, scalability, and relational (Krause et al., 2020).

3.4.1 VELOCITY OF BIG DATA

The rate of data generation in real-time scenarios is called the velocity of big data. Moreover, the velocity of big data is categorized into two main divisions:

- Velocity of data generation
- Velocity of managing the data (ie., data handling, data recording, data publishing)

3.4.2 VOLUME OF BIG DATA

Big data's unique attribute is to generate infinite data to handle, manage, and store. Moreover, the size of data is at least between terabytes and petabytes.

3.4.3 VERACITY OF BIG DATA

Veracity is defined as the ability to confirm data accuracy with a true value. To be exact, veracity shows the reliability of the data captured from the source. The quality of data is directly proportional to revenue generation. If data are hollow, the customer ignores the data process.

3.4.4 Variety of Data

Data variety is categorized into three types: structured data, unstructured data, and semi-structured data. Relational database managament systems (RDBMSs) are a tool generally used for handling structured data. But for unstructured and semi-structured data, special customized tools are essential to handle them.

3.4.5 Value of Data

The value of data could be increased only by analyzing and refining the exhaustive list of big data. Therefore, the value of data is directly proportional to the data refining processing time. Data quality can also be improved by multiple filters and the worth of information screened while processing it.

3.4.6 Variability of Data

Collected big data would not be available in specific formats or in the structure desired by analysts: raw data have to be assimilated and integrated into pieces, and eventually unstructured data can be transformed into structured data.

3.4.7 Scalability of Data

If big data is collected, the capacity to store the extensive data is an essential process. Therefore, data capture should not be restricted by limited memory. Moreover, data analysts must develop infrastructure for expanding memory.

3.4.8 Relational Data

The data collected have to be analyzed based on the correlational factors that link them. This can be effectively done by the meta-analysis process.

3.5 APPLICATIONS OF BIG DATA IN LOGISTICS

Based on statistics from the CEO of New Vantage Partners (Randy Bean), more than 91.6% of Fortune 1000 companies have invested heavily in big data and AI. This is done to avoid the establishment of competitors in the global landscape and because of "concern of disruption" in the commercial sphere (Osborne, 2020). Some of key companies that have invested in big data are: American Express, General Motors, Capital One, Ford, Mastercard, Johnson & Johnson, and UPS. The key applicable areas of big data use are effective planning of cargo circuits, warehouse automation, monitoring product quality, cargo mobility reports, and electrifying last-mile delivery.

3.5.1 Effective Planning of Cargo Circuits

Big data helped UPS drivers to reduce left turns, as they consumed more waiting time and increased idle fuel consumption and the potential for accidents. UPS management declared, "Because of left turns, trucks consumed 10 million gallons of fuel and emitted 20K tons of carbon dioxide."

Big data is a key tool to provide a better blueprint of resource utilization for last-mile delivery without supply surplus or lack of resources. Some of the constraints identified for logistic movers are as follows: dynamic weather conditions such as rainfall and snow avalanches, vehicles available for last mile delivery, diversion and shut-downs of highways and subways, and fuel cost surges.

Big data in logistics operates using the following micro-elements: real-time data on the transport fleet, driver accident behavior, road maintainance data, vehicle health and maintainance data, weather updates, temperature updates on cargo cabins, and so on.

3.5.2 WAREHOUSE AUTOMATION

Big data automation combined with business intelligence software helps automated package sorting centers arrange millions of packages without human interference. The robots operate with less power and within dedicated routes through connected servers. This helps logistics companies make the work simpler and faster with less manpower. Amazon hd introduced orange-coated KIVA robots for package mobility and sorting (Lebied, 2017). These robots required less investment and a minimal adoption time phase (less than a week).

3.5.3 MONITORING PRODUCT QUALITY

Customers' demands for faster delivery of fresh products impose heavy expectations on logistic movers. For example, if a logistic mover transports a perishable product (e.g., vegetables) from point A to point B (Evas, 2017) at a distance of 500 km, it requires temperature-controlled cabins. Moreover, the product temperature has to be reported to the customer, with an update frequency of 10 minutes to ensure satisfactory levels. This could be done by a Bluetooth chip (fixed in pallets) connected to the big data server.

3.5.4 CARGO MOBILITY REPORT

Cargo moving reports are seamlessly communicated to all stakeholders involved in product movement. Some of the key stakeholders actively participating in cargo mobility are the customer, logistic mover, third-party logistics, delivery person, and carrier. Moreover, the vehicles used by logistic agencies have sensors embedded, along with GPS-enabled smartphones, so these devices provide continuous updates on cargo mobility. These data are collected and stored in big data servers for predictive analysis, and hence future delivery frequency can be forecast with 95% accuracy.

3.5.5 ELECTRIFYING LAST-MILE DELIVERY

According to Mathew Winkebach, director of MIT Megacity logistics lab, last-mile analytics (using big data) quickens last-mile delivery further. It helps to store the delivery pattern in each area of the city through the GPS-enabled smartphones of delivery people and sensor-enabled delivery vehicles. This helps UPS companies operate at a non-peak delivery schedule with fewer distribution centers. Thus, it reduces the delivery cost and cancelled deliveries. Moreover, these data help identify vacant parking spots for delivery vehicles on each road.

3.6 ARTIFICIAL INTELLIGENCE

Artificial intelligence is the process of mimicking and simulating the natural intelligence of humans by machine learning and deep learning techniques. Machine learning is an educating process that primarily uses an automated learning process without human interference. Deep learning methodology is also a self-educating technology through unstructured data collection: text, images, videos, and others. The major difference between artificial intelligence and biological intelligence is consciousness and emotions. These two factors cannot be copied by intelligent machines.

3.6.1 Purpose of AI

The main significance of artificial intelligence (Meesseman et al., 2021) is to understand human speech, which includes the pitch and sharpness of the voice; to be able to play games with humans (chess, Go, and poker); to drive cars without human interference; and to execute military simulations. It enables machines to execute the following operations:

- Predicitive capability
- Operational warehouse process with robot (identifying packets within 0.2 sec)
- Computer vision (identifying freight damage and correcting it quickly)
- Better management of freight exceptions
- Resilient supply chain (Mangan et al., 2008)
- Visibility and control cost

3.6.2 AI Manages Redundant Data in a Manufacturing and Supply Chain—Case Study

Operational risk (Noble, 2021) in the supply chain could be effectively managed by a reliable and lean data approach (i.e., obsolete data have to be cleansed at a low frequency but with a high cost). Data cleansing could also be done using the discerning abilities of the analyst to build resilience into the supply chain. Artificial intelligence and machine learning make the data cleansing process more cost effective, rapid and user friendly to innovate more in business processes. Thereby the year-long cleansing process can be compressed to just a week in duration.

Ultra–AI-facilitated institutions are able to make the right decisions and develop larger intelligence. Moreover, AI processes connect the dots to provide a seamless process and to develop closer relationships between the partners of the supply chain to attain a common big picture. They also help to develop layers of visibility and enlarge the opportunity for data sharing within supply chain partners.

AI-enabled factories upgrade their day-to-day processes with multi-tasking intelligence. For example, management of an e-retail company in a European country wants to reduce the warehouse capacity in its offshore facilities, but this would take more than a month to a year if done manually.

Through AI, the process can be simplified through connected servers. While reducing warehouse capacity, the working capital used in the operating warehouse could also be optimized without much effort. Thus, AI reduces risk and boosts confidence in the company.

3.6.3 Significance of AI in Production and Supply Chain Visibility—Case Study

AI tools help to establish supply chain visibility (Maloney, 2021) primarily in two layers: transportation visibility and warehouse visibility. Transportation visibility means showing essential information (Radivojevic, 2016) to the user: inbound transportation data, outbound data, expected delivery time, and the type of freight moved. This allows companies to develop more resiliency when the freight is on the go. For example, a truck carrying goods to the port of Los Angeles is slowed down by unexpected traffic between the Los Angeles port and Long Beach. This disruption in the arrival schedule shifts other plans within the company, such as requiring more unloading manpower to sync with the time lag, delaying predictions of operations, affecting warehouse space release according to the arrival time, and informing customers about the postponed delivery. Therefore, supply chain optimization is adopted according to the delay in transportation.

The second layer of automation is warehouse automation, where AI-enabled robots are introduced to execute the volume of work. These robots also help warehouse operators to overcome labor-driven problems and delays in material handling. Moreover, process optimization is effectively done by these robots in warehouses. The robots help warehouse companies to visualize their operation and analyze it, for example, number of packets handled, quantity of incoming goods arriving per day, outgoing packets per day, damaged packets, and so on. A shortage of skilled operators is not a perpetual problem for these warehouses after the implementation of robots. This warehouse robot adoption required less than a week to install without altering the existing infrastructure.

3.7 MACHINE-TO-MACHINE TECHNOLOGY

M2M is a technology known for enhancing machine autonomy, interaction, and self-decision-making. Smart machines can communicate with each other through networks (Wi-Fi and satellite) by using equipment such as radio frequency identification (RFID), Bluetooth, and so on. M2M communication is initiated through a wired link (e.g., phone lines) and then furthered with wireless networks. M2M is exclusively used in warehouses, in logistics for replenishing orders, and to monitor the health of transport machines. The process involved in M2M is discussed in the following.

3.7.1 PROCESS OF M2M

The M2M process operates via three sub-element in different geographic locations, working in parallel through telemetric processes. M2M's three elements are a data capturer, cellular network, and remote computer. The data capturer (sensor, RFID) (Attaran, 2012) observes the temperature of a cabin, stock in the warehouse, fuel of a vehicle, or other data (Figure 3.1). Data capture has the ability to observe the quality of data from the host even under adverse conditions. It has a data filter to avoid noise but extract signals.

The second sub-element is a cellular network, which acts as a link between the data capturer and the remote computer. It uses telemetry processes (Shea, 2019), where data collection happens at the remote end through a network. Transmission is then initiated with telephonic lines and then with dedicated radio waves allotted by internet and seamless connectivity modes (optic fibers).

The third sub-element is a remote super computer, which can process the data and facilitate machines to make decisions autonomously. For example, a connected home's heater could be triggered automatically based on the atmospheric temperature of the home (without human interference).

3.7.2 APPLICATIONS OF M2M

1. M2M is used for telemedicine (remote health monitoring) of patients by doctors if a disease is highly infectious to others and hospital bed scarcity prevails.
2. By using M2M, machines can monitor smart home equipment such as lighting systems and heating and ventilation systems.
3. M2M machines are used for harvesting energy and to track the energy usage of customers by using customers' smart meters.

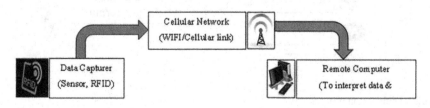

FIGURE 3.1 M2M architecture.

3.7.3 Features of M2M

- M2M demands minimal power consumption for effective system operation.
- Effective time control mechanisms allow machines to send or receive data at specific time intervals.
- The network strength also allows machines to have visual tracing of remote locations.
- Alerts can be enabled for geographically specific data to identify data locations.

3.7.4 M2M in Smart Logistics

M2M communications require (long-term evolution; LTE) seamless connectivity by generating an extensive number of nodes and patterns in short time periods. The data communicated through M2M are temperature, pressure, humidity, and so on, which were extracted from surrounding zones. The exponential rise (Mehmood et al., 2016) of vehicles on roads provided a greater opportunity to extend the market space of M2M. Infrastructure (roads, bridges) and dynamic objects (cars and trucks) interact through M2M technology. M2M uses RFID and near-field communication (NFC) to transmit data from stationary infrastructures and dynamic objects to the nearest logistic control towers to track the health of cargo transit. It also detects the location of vehicles (once per 10 minutes) and reports the updated status to customers frequently.

In addition to that, containers travelling through international waters have a high probability of being damage, such as through piracy, thefts, delivery lag time, and ship sinkings. Therefore, these containers have to be constantly monitored to ensure customer satisfaction. Data that can be observed from the container cargo include fuel consumption, fleet safety, position of containers, and product safety. It can also provide more control (Osborne, 2020) over the cargo, higher resource management, and more cost effectiveness.

Moreover, M2M technology in warehouses tracks the inventory level in each rack and pallet and updates the stakeholders of the warehouse on this information. Based on the information, the stock of warehouses can be handled responsibly. These technologies also help warehouse investors to operate within lean margins and provide rapid response to customer demands. Eventually, each M2M-enabled warehouse will need low manpower to operate it, without much capital investment.

3.8 DIGITIZATION

The process of transforming any form of information (image, voice, text, sound) into a digital format is known as digitization. The transformed format (digital) carries a series of binary numbers. Therefore, in short, digitization is the process of transforming analog into digital format. The digitization process (Figure 3.2) enables data to have more speed and accuracy, with no or less degradation. Digitization increases productivity and efficiency.

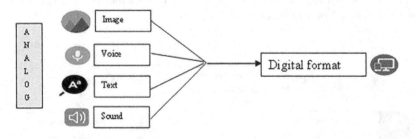

FIGURE 3.2   Digitization process.

3.8.1 Benefits of Digitization

- Digitization helps store, manipulate, and reproduce data in any number of iterations.
- Digitization helps control data remotely (Sen Gupta, 2020), with greater level of accuracy.
- Digitization also helps data managers establish uniform governance.
- Digitization transforms data control rights from localized to centralized control. Hence, it increases data visibility in all layers.
- Moreover, data is easy to filter and analyze rapidly.

3.8.2 Significance of Digitization

- Digitization permits a higher level of automation in business.
- Digitization provides more intelligence in the business process (i.e., artificial intelligence and big data can be enabled if digitization is implemented).
- Digitization also permits more social collaboration with data and a common platform to communicate (e.g., a team working on data).
- It also permits data to have more digital revenue streams.

3.8.3 Digitized Logistics Evolution—A Study

In recent decades, the consumer domain has experienced transformation from mere digital transformation to the Internet of Things (IoT) and services. It demands intelligent platforms to connect people (consumer, manufacturer, packers), data, and machines. Moreover, it places high pressure on logistic movers to have high speed, flexibility, and more control over the process. Thus, the logistic process was termed "Logistics 4.0," which operates with IoT sensors, digital transmitters, and so on.

Logistics 4.0 is defined as the process of connecting the processes of inbound and outbound logistic operations of movers and production zones with customers. Logistics 4.0 allows sensors, cameras, and warehouse robots to provide more layers of visibility and better tracking. These input devices also assist users in optimized decision making.

The transformation of analog processes to digital modes adds more significance, such as value and transparency in logistics. Prior to the digitization of logistics, compatible business models had to be adopted in all layers of suppliers who operated the logistics network. This prerequisite was essential for effective logistics digitization. Moreover, chartered cargo will become more intelligent to hire suitable transport (full truck load; FTL/less than truckload; LTL) through big data technology in the future. Logistics 4.0 could enhance the extent of in-house automation to cross-company automation by overriding its limits and developing material flow optimization.

3.8.4 Applications of Logistics 4.0

A forecast of digital adoption (Scher, 2019) and automation has boosted productivity by 30%. It has motivated leading logistic players to migrate from hardcopy to digital documents and paperless transactions. Moreover, prior to digital migration, compatible and intelligent platforms had to be developed between suppliers. Some case studies are given in the following.

The DHL app (My Way) in Sweden allows customers to collect parcels for delivery to neighbors when they return to their homes from offices; Uber Rush, a new wing introduced in the United States, helps customers pick the right truck for their parcels with a low and affordable cost. Moreover, this wing operates on digital orders. Zalando, an online shoe retailer, effectively collects used shoes from customers for remanufacturing through digital modes.

3.9 CONCLUSION

This chapter examines the concepts of Industry 4.0 and its influence on logistics and the supply chain. For classical logistics without automated machines and GPS networks, advancements like same-day delivery and trucks with shared space (for multiple customers) were impossible tasks. But the advent of new technologies helps people by increasing customer comfort and delivering products on the same day of the order itself, with minimal defects. In addition to that, logistic planners can visualize and simulate future delivery orders based on past data, and this can help them prepare with additional facilities such as fulfillment centers and delivery trucks. Therefore, orders that satisfy customers have been transformed into orders that delight customers.

Most of the leading corporate players after the pandemic have invested (30–50%) in digital infrastructure as a part of business sustenance. Some of these are bot-enabled big data technology, unmanned drones for last-mile delivery using AI systems and M2M communications, and avoiding customer signature by One Time Password (OTP)-enabled acknowledgements. In addition, manual ledgers can be translated into digital ledgers to avoid physical contact. DHL has transformed its physical documents (customs clearance, invoices) into digital documents for easy visibility and to reduce human contact. Moreover, customers are offered big data–enabled Google Lens and virtual reality tools to visualize product information. Industry 4.0 tools have extended their utility across all domains of the industry and service sectors. Boeing trains their pilots on handling aircraft using AI-enabled simulators, whereas sports car maker Lamborghini trains owners of its high-valued cars with simulators before they become users. In the United States, California-based police troops are trained with simulators to protect civilians from mass shootings. These applications demonstrate the scale of the reach across multiple customer sectors irrespective of geography and age.

REFERENCES

Attaran, M. (2012). Critical Success Factors and Challenges of Implementing RFID in Supply Chain Management. *Journal of Supply Chain and Operations Management*, 10(1), 114–167.

Attaran, M. (2017). Cloud Computing Technology: Leveraging the Power of the Internet to Improve Business Performance. *Journal of International Technology and Information Management*, 26(1), 112–137.

Buyukozkan, G. & F. Gocer. (2018). Digital Supply Chain: Literature Review and a Proposed Framework for Future Research. *Computers in Industry*, 97, 157–177. https://doi.org/10.1016/j.compind.2018.02.010.

Charlie Osborne. (2020). *Fortune 1000 to 'Urgently' Invest in Big Data, AI in 2019 in Fear of Digital Rivals*. www.zdnet.com/article/fortune-1000-to-urgently-invest-in-big-data-ai-in-2019-in-fear-of-digital-rivals/

Chase, C. (2019). How the Digital Economy Is Impacting the Supply Chain. *Journal of Business Forecasting*, 38(2), 16–20.

David Maloney. (n.d.). *Longbow Advantage on Supply Chain Visibility (No. 21)*. Retrieved May 28, 2021, from www.dcvelocity.com/articles/51344-the-logistics-matters-podcast-alex-wakefield-of-longbow-advantage-on-supply-chain-visibility-season-2-episode-21

Digital Report. (2020). *What Is Big Data? Introduction, Types, Characteristics, Example*. www.guru99.com/what-is-big-data.html

Ernest Krause et al. (2020). *Big Data*. https://en.wikipedia.org/wiki/Big_data

Evas, T. (2017). Public Consultation on Robotics and Artificial Intelligence-First Results of Public Consultation (No. 2017). *European Parliamentary Research Service*. www.epthinktank.eu

Geisberger, E. & M. Broy. (2012). *Agenda CPS—Integrierte Forschungsagenda Cyber-Physical Systems*. Springer, Munich.

Jeske, M., M. Gruner & F. Weib (2013). *Big Data in Logistics*. www.dhl.com

Jonas Scher. (2019). *What Is Logistics 4.0? Everything You Need to Know about Digitization & Logistics*. MM Machinen International. www.maschinenmarkt.international/what-is-logistics-40-everything-you-need-to-know-about-digitization-logistics-a-876611/

Josh Fruhlinger. (2020). *What Is IoT? The Internet of Things Explained*. www.networkworld.com/article/3207535/what-is-iot-the-internet-of-things-explained.html

Kearney, A. T. (2015). *Digital Supply Chains: Increasingly Critical for Competitive Edge*. www.atkearney.com/documents/20152/435077/Digital%2BSupply%2BChains.pdf/82bf637e-bfa9-5922-ce03-866b7b17a492

Kuckelhaus, M. & G. Chung. (2018). *Logistics Trend Radar, DHL Customer Solutions & Innovation.* Germany. www.dhl.com

Laurent Meesseman et al. (2021). *Artificial Intelligence.* https://en.wikipedia.org/wiki/Artificial_intelligence

Mangan, J., C. Lalwani & T. Butcher. (2008). *Global Logistics and Supply Chain Management.* John Wiley & Sons, New Jersey, NJ.

Mark Sen Gupta. (2020). *What Is Digitization, Digitalization, and Digital Transformation?* [Industry trends]. www.arcweb.com/blog/what-digitization-digitalization-digital-transformation

Mona Lebied. (2017). *5 Examples of How Big Data in Logistics Can Transform the Supply Chain (Business Intelligence).* www.datapine.com/blog/how-big-data-logistics-transform-supply-chain/

Paul Noble. (2021). *How AI Tackles the Problem of 'Dirty' Data.* www.supplychainbrain.com/blogs/1-think-tank/post/33100-how-ai-tackles-the-problem-of-dirty-data

Radivojevic, G. (2016). *Information Management in Logistics. Faculty of Transport and Traffic Engineering,* University of Belgrade.

Sharon Shea. (2019). *Machine-to-Machine (M2M).* https://internetofthingsagenda.techtarget.com/definition/machine-to-machine-M2M

Vaidya, S., P. Ambad & S. Bhosle. (2018). Industry 4.0—A Glimpse. *Procedia Manufacturing,* 20, 233–238.

Yasir Mehmood et al. (2016). M2M Potentials in Logistics and Transportation Industry. *Logistic Research Journal,* 9(15), 1–11. https://doi.org/10.1007/s12159-016-0142-y

4 Digital Twin–Based Smart Manufacturing—Concept and Applications

Vishal Ashok Wankhede, Rohit Agrawal, and S. Vinodh

CONTENTS

4.1 Introduction 31
4.2 Application Scenarios of Digital Twins 32
    4.2.1 Digital Twins for Production Shop Floors 32
    4.2.2 Digital Twins for Manufacturing Systems 33
    4.2.3 Digital Twins for Assembly Operation and Inspection 33
    4.2.4 Digital Twins for Smart Manufacturing 33
    4.2.5 Digital Twin Integration with I4.0 Technologies 34
    4.2.6 Digital Twin–Based Frameworks 34
4.3 Challenges Concerning the Suitability of Digital Twins....................................34
4.4 Conclusions and Future Research Directions 35
References...35

4.1 INTRODUCTION

The term "industrial revolution" showed a shift in the manufacturing process. It began in the 16th century when wealthy merchants gathered laborers to produce fabrics at home. Creating the first steam engine to generate power from steam was considered the first industrial revolution (Vinodh et al., 2021). The assembly line concept was established with time, which shortened production lead time and was called the second industrial revolution. The introduction of computers in industry led to computer-integrated production. The third industrial revolution began with adopting information technologies and computers, which replaced humans with semi-automated robots. Industry 4.0 (I4.0), also known as the fourth industrial revolution, started with the digitization of the manufacturing system by combining virtual and physical worlds. Its goal is to create intelligent machines with intelligent components that include a sensor system for data collection and an actuator system for physical process control, both of which happen in real time.

With the growing use of I4.0 technologies such as artificial intelligence (AI), the Internet of Things (IoT), cloud computing, and big data, several initiatives for distributed manufacturing are being explored (Wankhede and Vinodh, 2021). One common goal of such strategies is to develop intelligent manufacturing with more sustainability, intelligence, and customization. Interaction and integration among the virtual and real worlds of production is a necessary prerequisite.

In recent years, academicians and industrial practitioners have become more interested in digital twins (DTs) as a possible technological way of achieving cyber-physical system integration, interaction, and fusion. DTs may accurately duplicate a physical item or process in the virtual world by high-fidelity modeling, interaction based on real time, and data fusion, allowing for more efficient monitoring, control, and prediction of the product or machine throughout its entire life. As described by NASA, "DT is an integrated, multiphysics, multi-scale, probabilistic simulation of an

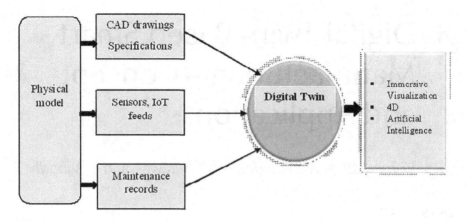

FIGURE 4.1 Digital twin representation.

as-built vehicle or system that uses the best available physical models, sensor updates, and fleet history to mirror the life of its flying twin" (Roy et al., 2020).

The goal of a DT is to create a virtual and digital model of a physical thing that can mimic its life and make helpful predictions about it. The current chapter discusses the significance of DTs and gives a detailed overview of their applications, along with the possibility of incorporating them into specific industrial processes. For better understanding of DTs, a schematic diagram is presented in Figure 4.1. This chapter presents the concept of DTs by discussing various application scenarios through state-of-the-art review. In addition, challenges for the suitability of DTs are discussed.

4.2 APPLICATION SCENARIOS OF DIGITAL TWINS

4.2.1 DIGITAL TWINS FOR PRODUCTION SHOP FLOORS

Digital twins possess significant importance in manufacturing due to their capability to detect bottlenecks, optimize the manufacturing process, and simulate situations to predict operational performance. Raza et al. (2020) developed a DT model to replicate real-time production line processes for production assembly. The authors implemented a graphical/digital model of a cyber-physical factory for I4.0, allowing scheduling of orders and identification of the product configurations that represent the workings of the cyber-physical factory in the digital world. The developed digital model offered improved services and a route map for understanding completely operational DTs. The authors comprehensively depicted the application of the DT in the representation of assembly-line applications. Fan et al. (2021) presented a DT visualization-based architecture for responsive manufacturing systems. The authors investigated the DT cyber-physical modeling of multi-source heterogeneous information and explored the DT scenario of 3D visualized human–machine interaction within the proposed architecture. Guo et al. (2021) illustrated a DT-based layout optimization technique for the manufacturing workshop. The authors considered the workshop-related layout problem and solved it using twin data fusion, physical and information interaction fusion, and optimization and data analysis. The authors established a DT-based workshop sub-framework and optimized workshop portioning using simulation analysis. Further, equipment layout decisions were made using real-time collection of twin data. The proposed framework was adopted in the welding workstation and increased production capacity by 29.4%.

4.2.2 Digital Twins for Manufacturing Systems

Cimino et al. (2019) analyzed the degree of adoption of DT systems in the manufacturing paradigm. The authors explored various DT applications with respect to manufacturing and the associated service provided by DTs in manufacturing sectors. The study further deployed DT technology to monitor energy consumption associated with the manufacturing line process. Roy et al. (2020) reviewed several DTs along with their application domains. The authors developed schematic images of few DTs from various fields. Further, the model was explained using a case study of developing a DT for an advanced manufacturing process called friction stir welding. Lugaresi and Matta (2021) proposed a technique to discover manufacturing systems and create enough DTs automatically. The significant characteristics of the production system were automatically recovered from data logs. The developed technique was applied to a real manufacturing line and on two test cases. The experimental results proved the effectiveness of creating digital models that correctly estimated the system performance.

4.2.3 Digital Twins for Assembly Operation and Inspection

A DT prototype comprising production and operation stages was suggested for a shop floor involved in assembly operation (Roy et al., 2020). The construction stage includes creating IoT networks in the production shop floor for gathering digital data of manufacturing resources, namely tags, bar codes, and so on, to allow real-time information flow. Polini and Corrado (2020) introduced a DT tool to support lightweight composite material assembly design. The developed tool reproduced the physical manufacturing process virtually to create geometrical deviations of the manufacturing part. Further, testing of the set of manufactured parts was carried out and estimated the variability. The developed virtual tool exchanged data with physical processes and updated geometrical deviations of in-process components. Židek et al. (2020) created a DT model of an automized line and belt conveyor-based experimental assembly system for quality production inspection. The assembly system was situated in a fixture with IoT devices and ultra–high frequency tags to identify position and status. The generated DT model helped in the inspection and identification of manufactured parts.

4.2.4 Digital Twins for Smart Manufacturing

The concept of a DT has the potential to be a key component of smart manufacturing. However, there is still a lot of misunderstanding regarding the concept and how it may be applied in real-world production processes, particularly among small and medium-sized businesses (Shao and Helu, 2020). The authors summarized various perspectives on DTs that have been reported to highlight the main qualities for practical applications. Shao et al. (2019) discussed the DT concept for manufacturing research and practice for the simulation community. A panel discussion was done to obtain preliminary thoughts on definitions, concepts, challenges, relevant standard activities, implementations, and future directions. The discussion also presented several research aspects of DTs for manufacturing. Evangeline and Anandhakumar (2020) discussed the DT concept and its application for smart manufacturing. The authors explained the significance of the DT and its integration with analytics and actuators. Moreover, DT business values were also elaborated on. Zhang et al. (2021) proposed a DT-based smart manufacturing cell architecture to deal with fluctuations in the manufacturing environment. The developed architecture possesses the ability to analyze, sense, and make decisions to assist autonomous manufacturing. Resource allocation in distributed manufacturing has been aided by an integrated design, as Wang et al. (2021) proposed. The authors proposed a service model based on a DT for continuous monitoring and control of distributed manufacturing. Further, the author presented a case study to validate the efficiency of the proposed service model. Leng et al. (2021a) proposed a DT model for the open architecture of manufacturing systems, enabling remote monitoring of manufacturing systems. The proposed model was applicable for a flow-type manufacturing system. The proposed model helped industrial practitioners to find hidden errors that can be overcome before the implementation phase. Leng et al. (2021b) presented a review on adopting DTs in the smart manufacturing system. Various technologies that help smooth DT adoption in manufacturing paradigms such as IIOT, data analytics, blockchain, and AR/VR were analyzed. The authors recommended that DTs will significantly improve the smart manufacturing system.

4.2.5 DIGITAL TWIN INTEGRATION WITH I4.0 TECHNOLOGIES

In the era of I4.0, additive manufacturing (AM) played a critical role in modern manufacturing. Several AM challenges, such as process modeling, monitoring, and control, may now be addressed with DTs (Zhang et al., 2020). They can aid in a deeper understanding of the functions of various manufacturing factors and their sensitivity to the quality of the product. Furthermore, the system can give feedback data for active manufacturing process control. With input parameters, DT technology could visualize the entire AM process, and Key Performance Index (KPIs) of the system could be predicted promptly and precisely (Zhang et al., 2020). DTs helped a system obtain greater connectedness and flexibility and increased intelligence (Rojek et al., 2021). Furthermore, manufacturing efficiency and quality of the goods produced are improved, lowering costs and improving revenues. Aivaliotis et al. (2019) presented a model based on a DT to evaluate the useful life remaining of any equipment. The data were collected by deploying various sensors in the equipment and using a DT model for simulation. The simulation results helped in predicting the remaining life of the equipment. The authors presented a case study to validate the proposed model. DTs, which are living virtual duplicates of real processes, will enhance AI capabilities. DTs will be able to supervise processes autonomously and will be able to be questioned by industry practitioners to determine the best processing path for every given product (Gunasegaram et al., 2021). The authors emphasized the importance of DTs in AM and the requirements to develop high-end models for AM processes to improve AI capabilities. The authors identified technical roadblocks to DT system development, such as those originating from the models' multi-scale challenges integrating subprocess models and a lack of experimental data.

4.2.6 DIGITAL TWIN–BASED FRAMEWORKS

Digital twin–based architecture/frameworks provide real-time and service-based enabled infrastructure for horizontal and vertical integration. Redelinghuys et al. (2020) presented a DT-based architecture enabling sharing of information and data between a remote simulation and physical twin. The architecture encompassed several layers: IoT gateway layer, local data layer, emulations and simulation layer, and cloud-based databases layer. The architecture was employed in legacy and new production facilities with minimum disturbance of existing installations. The architecture was evaluated using a small physical manufacturing system component. In the future, the developed six-layer architecture could be extended to accommodate clusters of DTs. Guo et al. (2020) proposed a DT, additive manufacturing, and blockchain-based personalized production framework with respect to Industry 4.0. The developed framework was discussed from the perspectives of environmental and social effects, a customer-centric business model, and data ownership challenges. The quantitative analysis of the blockchain, additive manufacturing, and DT with respect to personalized production performance was suggested. Moreover, to obtain coordination between the manufacturer and customer, sharing and collaboration mechanisms were suggested.

4.3 CHALLENGES CONCERNING THE SUITABILITY OF DIGITAL TWINS

DTs are one of the disruptive technologies of I4.0 and have been recognized for the continuous re-optimization of logistics operations and manufacturing. DTs possess several benefits, such as increasing productivity, enhancing performance, reducing operational costs, and changing how predictive maintenance is done. Despite these advantages, DTs hold various challenges in manufacturing (Shao et al., 2019), as follows:

- Creating the virtual model consisting of numerous sub-modules
- Interfaces for standardized information exchange
- Effective design of information flow
- Time and cost for digital twin development
- Complex system engineering
- Standardization
- Scalability and supply chain
- Sector servitization
- Information sharing
- Digital security
- Data privacy concerns
- Data consistency
- Lack of awareness of the digital twin
- User interaction
- Lack of skilled competencies

4.4 CONCLUSIONS AND FUTURE RESEARCH DIRECTIONS

One strategy of Industry 4.0, which deals with operating on the virtualization principle, is the DT technology. This chapter presented the concept of DTs by discussing various application scenarios through state-of-the-art review. In addition, challenges concerning the suitability of DTs were discussed. The chapter would help industry practitioners and managers understand the significance of digital twins in manufacturing and motivate them to improve manufacturing performance. Based on the review of previous studies, the following future research directions are suggested for researchers:

- Case studies pertaining to combining DT technology with optimization of workshop layout.
- Analyzing the degree of adoption of DT systems in a manufacturing paradigm.
- Examining the identified DT challenges for smooth adoption of DTs in manufacturing industries.
- Generating the DT model for inspection and identification of manufactured parts.
- Utilizing DTs to integrate more communication methods and devices in the system to realize the system's function.
- Identification and analysis of the technical roadblocks to DT system development.
- Obtaining coordination between manufacturer and customer and the development of sharing and collaboration mechanisms using DTs.
- Generating DTs for cost control and efficiency improvement of the shared manufacturing.

REFERENCES

Aivaliotis, P., Georgoulias, K., & Chryssolouris, G. (2019). The use of digital twin for predictive maintenance in manufacturing. *International Journal of Computer Integrated Manufacturing, 32*(11), 1067–1080.

Cimino, C., Negri, E., & Fumagalli, L. (2019). Review of digital twin applications in manufacturing. *Computers in Industry, 113*, 103130.

Evangeline, P., & Anandhakumar, P. (2020). Digital twin technology for "smart manufacturing." In *Advances in Computers* (1st ed., Vol. 117, Issue 1). Elsevier Inc. https://doi.org/10.1016/bs.adcom.2019.10.009

Fan, Y., Yang, J., Chen, J., Hu, P., Wang, X., Xu, J., & Zhou, B. (2021). A digital-twin visualized architecture for flexible manufacturing system. *Journal of Manufacturing Systems*, *60*(January), 176–201. https://doi.org/10.1016/j.jmsy.2021.05.010

Gunasegaram, D. R., Murphy, A. B., Barnard, A., DebRoy, T., Matthews, M. J., Ladani, L., & Gu, D. (2021). Towards developing multiscale-multiphysics models and their surrogates for digital twins of metal additive manufacturing. *Additive Manufacturing*, 102089.

Guo, D., Ling, S., Li, H., Ao, D., Zhang, T., Rong, Y., & Huang, G. Q. (2020). A framework for personalized production based on digital twin, blockchain and additive manufacturing in the context of Industry 4.0. *IEEE International Conference on Automation Science and Engineering*, 2020-August (September), 1181–1186. https://ieeexplore.ieee.org/abstract/document/9216732?casa_token=l_eyZ-frD7IAAAAA:_9YhQ-MyBDCMA4mlrExxOK1_IbA7Yk571o92H964pmaPIcmcxFiuXh4hRq0QMdxMjZ5Xk0H4aB3P. https://doi.org/10.1109/CASE48305.2020.9216732

Guo, H., Zhu, Y., Zhang, Y., Ren, Y., Chen, M., & Zhang, R. (2021). A digital twin-based layout optimization method for discrete manufacturing workshop. *International Journal of Advanced Manufacturing Technology*, 1307–1318. https://doi.org/10.1007/s00170-020-06568-0

Leng, J., Wang, D., Shen, W., Li, X., Liu, Q., & Chen, X. (2021b). Digital twins-based smart manufacturing system design in Industry 4.0: A review. *Journal of Manufacturing Systems*, *60*, 119–137.

Leng, J., Zhou, M., Xiao, Y., Zhang, H., Liu, Q., Shen, W., . . . & Li, L. (2021a). Digital twins-based remote semi-physical commissioning of flow-type smart manufacturing systems. *Journal of Cleaner Production*, *306*, 127278.

Lugaresi, G., & Matta, A. (2021). Automated manufacturing system discovery and digital twin generation. *Journal of Manufacturing Systems*, *59*(January), 51–66. https://doi.org/10.1016/j.jmsy.2021.01.005

Polini, W., & Corrado, A. (2020). Digital twin of composite assembly manufacturing process. *International Journal of Production Research*, *58*(17), 5238–5252.

Raza, M., Kumar, P. M., Hung, D. V., Davis, W., Nguyen, H., & Trestian, R. (2020). A digital twin framework for Industry 4.0 enabling next-gen manufacturing. *ICITM 2020–2020 9th International Conference on Industrial Technology and Management*, February, 73–77. https://doi.org/10.1109/ICITM48982.2020.9080395

Redelinghuys, A. J. H., Basson, A. H., & Kruger, K. (2020). A six-layer architecture for the digital twin: A manufacturing case study implementation. *Journal of Intelligent Manufacturing*, *31*(6), 1383–1402.

Rojek, I., Mikołajewski, D., & Dostatni, E. (2021). Digital twins in product lifecycle for sustainability in manufacturing and maintenance. *Applied Sciences*, *11*(1), 31.

Roy, R. B., Mishra, D., Pal, S. K., Chakravarty, T., Panda, S., Chandra, M. G., Pal, A., Misra, P., Chakravarty, D., & Misra, S. (2020). Digital twin: Current scenario and a case study on a manufacturing process. *International Journal of Advanced Manufacturing Technology*, *107*(9–10), 3691–3714.

Shao, G., & Helu, M. (2020). Framework for a digital twin in manufacturing: Scope and requirements. *Manufacturing Letters*, *24*, 105–107.

Shao, G., Jain, S., Laroque, C., Lee, L. H., Lendermann, P., & Rose, O. (2019). Digital twin for smart manufacturing: The simulation aspect. *Proceedings—Winter Simulation Conference*, 2019-December (Bolton 2016), 2085–2098. https://doi.org/10.1109/WSC40007.2019.9004659

Vinodh, S., Antony, J., Agrawal, R., & Douglas, J. A. (2021). Integration of continuous improvement strategies with Industry 4.0: A systematic review and agenda for further research. *The TQM Journal*, *33*(2), 441–472.

Wang, G., Zhang, G., Guo, X., & Zhang, Y. (2021). Digital twin-driven service model and optimal allocation of manufacturing resources in shared manufacturing. *Journal of Manufacturing Systems*, *59*, 165–179.

Wankhede, V. A., & Vinodh, S. (2021). Analysis of Industry 4.0 challenges using best worst method: A case study. *Computers & Industrial Engineering*, 107487.

Zhang, J., Li, P., & Luo, L. (2021). Digital twin-based smart manufacturing cell: Application case, system architecture, and implementation. *Journal of Physics: Conference Series*, *1884*(1). https://doi.org/10.1088/1742-6596/1884/1/012017

Zhang, L., Chen, X., Zhou, W., Cheng, T., Chen, L., Guo, Z., . . . & Lu, L. (2020). Digital twins for additive manufacturing: A state-of-the-art review. *Applied Sciences*, *10*(23), 8350.

Židek, K., Pitel, J., Adámek, M., Lazorík, P., & Hošovský, A. (2020). Digital twin of experimental smart manufacturing assembly system for Industry 4.0 concept. *Sustainability (Switzerland)*, *12*(9), 1–16.

5 Applications of Augmented and Virtual Reality in Contemporary Manufacturing Organisations

Santosh Kumar, Vijaya Kumar, and Sakthi Balan Ganapathy

CONTENTS

5.1 Fourth Industrial Revolution—An Introduction...37
5.2 The Goal of Industrial Revolution 4.0 38
5.3 Augmented Reality and Virtual Reality 38
    5.3.1 Advantages of AR/VR...39
5.4 Augmented Reality in Industry 4.0 39
5.5 Uses of Augmented Reality Applications in Industry 4.0...................................39
    5.5.1 Complex Assembly 39
    5.5.2 Maintenance 40
    5.5.3 Expert Support 40
    5.5.4 Virtual Reality in Industry 4.0..40
5.6 Uses of Virtual Reality Applications in Industry 4.0..40
    5.6.1 Training..40
    5.6.2 Factory Planning 40
    5.6.3 Inspection 40
5.7 AR and VR Used for Changes in Manufacturing Systems 41
5.8 AR/VR Applications in Other Manufacturing Areas 42
    5.8.1 Design and Prototyping..42
    5.8.2 Inventory Management...42
    5.8.3 Prevention of Accidents and Disruptions 43
    5.8.4 Inspection and Maintenance..43
5.9 AR/VR Applications in Various Areas 44
    5.9.1 Healthcare 44
    5.9.2 Education..44
    5.9.3 Culture and Tourism..45
    5.9.4 Gaming...45
    5.9.5 Real Estate...45
    5.9.6 Emergency Management...45
    5.9.7 Mental Health Services 46
5.10 Augmented and Virtual Reality during COVID...46
References..46

5.1 FOURTH INDUSTRIAL REVOLUTION—AN INTRODUCTION

In the history of manufacturing, there are numerous revolutions, and the next one is currently underway, called 'connectivity.' The complete digitisation of manufacturing plays a major role in the

DOI: 10.1201/9781003186670-5

manufacturing process. In the future it will be more important than ever to operate quickly, flexibly, and cost effectively. Lead times will become shorter, so delivery must be faster. Companies in the future must align their organisations much more rigorously to meet customer demands. This can only be achieved if companies start to digitise, automate, and above all interconnect all processes not only within the company but also the entire value chain. This means the convergence of various areas. The customers and business partners are part of the total process. The goal is to make all relevant data available any time and be able to control the whole network of value creation in real time from product planning and development all the way up to production logistics. Data security must be ensured all the times. Synchronising digital and physical value streams not only increases productivity of companies but also their efficiency, quality, and innovative capacity. Industry 4.0 knowledge and technology make the future together secure and cost effective with quality and efficiency.

The intelligent automation and incorporation of new technology into the business value chain is the fundamental feature of Industry 4.0, or the fourth industrial revolution. This is a digital change that revolutionises the company by radically altering not just systems and processes but also management styles, business models, and the workforce [1].

The concept of Industry 4.0 is a new way of organising production methods. The revolution supports the creation of a "smart factory" that integrates physical and digital systems with the purpose of "mass personalization" and "faster product development" [2].

The objective is to set up "smart factories" capable of making production more adaptable and allocating resources more efficiently, paving the path for a new industrial revolution. The Internet of Things and cyber-physical systems are the technological foundation [3].

5.2 THE GOAL OF INDUSTRIAL REVOLUTION 4.0

Manufacturing firms have faced numerous obstacles in recent years, including those linked to developing innovative goods, variable demand, and changing customer and supplier requirements, all of which involve new technical roadmaps and interventions in manufacturing systems. Intelligent and qualified operators, known as smart operators, perform work with the assistance of machines, communicate with collaborative robots and advanced systems, and employ technology like wearable devices and augmented and virtual reality. Augmented and virtual reality (A/V) can be used for workforce training, and they should be able to interact with a human workforce effectively. The goal is to create and implement integrated VR and AR manufacturing systems that can improve production processes, as well as product and process development, resulting in shorter lead times, lower costs, and higher quality. The ultimate goal is to construct a system that is as good as, if not better and more efficient than, the real world.

5.3 AUGMENTED REALITY AND VIRTUAL REALITY

Augmented reality (AR) is a technology that augments the real-world factory environment by superimposing information and digital data in real time in the field of view (e.g. headphones, smart phones, tablets, or AR space projectors).

Virtual reality (VR) instead is a computer-simulated multimedia reality that can digitally replicate a design environment and allow an intelligent operator to interact with any presence within (e.g., a product, a machine tool, a robot, a line production, a factory) with reduced risk, real-time feedback, and cost minimisation.

Through a purely cognitive style of contact, AR and VR involve numerous firm stakeholders such as managers, employees, maintenance operators, production operators, and logistics operators. The biggest distinction in how these technologies are used is in the application fields. The augmented operator is a real-life operator who employs this technology to assist him or her in his daily work tasks, with virtual reality serving as a great training environment [4].

5.3.1 Advantages of AR/VR

The advantages that AR technology can provide are

1. Faster cycle times
2. Shorter completion times
3. Improved reliability
4. Low error rates
5. Shorter learning curve
6. Increased perceived efficiency in performing the task
7. Improved health and safety
8. Increased employee engagement, motivation, and flexibility
9. Employability of operators

These advantages are dependent on AR's improvement of cognitive abilities, such as improved problem-solving, decision-making, memory, and understanding, along with reduced cognitive load, reduced mental workload associated with a task because only task-relevant information is displayed, and reduced cognitive distance because the information space and the physical space coincide.

Operators who utilise this technology, however, must have abilities in using new digital interfaces and interacting with holograms, as well as the capacity to read and analyse real-time data, make quick decisions, and solve complex problems because the relevant information is available.

Virtual reality has proven to be a safe environment that can be utilised with little or no actual prototypes, is easy to alter for new models or equipment, and can be used with little supervision. Virtual reality has also been demonstrated to increase student engagement in learning when compared to traditional teaching methods, and the costs of errors in virtual reality are low when compared to the possible costs in the real world.

5.4 AUGMENTED REALITY IN INDUSTRY 4.0

Augmented reality identifies a set of technologies which allow viewing the real-world environment in an augmented or enhanced way through computer-generated graphics. This visual aspect in the physical environment is enhanced by using various devices. Augmented reality has varied applications, but manufacturing is one domain which can be most attractive for the world of augmented reality.

Concerned with various processes to transform raw materials into finished goods by adding value to them, augmented reality can be a real game changer. This is because real-time information is needed at the various stages of product life cycle. From design to prototyping, to production and assembly and maintenance, each stage has its own sets of challenges. Augmented reality can be a boon in these complex processes, as it is capable of simulating, assisting, and improving the processes even before they are carried out. A lot of manufacturing units are now open to this idea of utilising augmented reality, simulating processes, reducing downtime, and streamlining operations.

5.5 USES OF AUGMENTED REALITY APPLICATIONS IN INDUSTRY 4.0

5.5.1 Complex Assembly

Modern manufacturing involves assembling hundreds of complex components in a short time with precision. Augmented reality can help in these complex assemblies. The work documents are generally in readable format, which is difficult to carry on. Augmented reality can help to bring them to life in a video. They are made glanceable in the field of view, which is hands free and voice controlled. The instructions are broken down and video can be added. All this can be seen through AR glasses while workers keep their hands on the task.

5.5.2 Maintenance

After assembly, maintenance is another aspect where augmented reality can play a crucial role. Currently, most workers confirm maintenance manually using a manual. This process can be time consuming and not 100% error free. Many companies have been developing technology to support maintenance using AR based on 3D models. This would enable users to confirm that the order of inspection is followed, and inspection results can be added. More specifically, the machine's status can be checked only via glancing at it through AR glasses, which can be a powerful maintenance tool.

5.5.3 Expert Support

In the event of a disturbed manufacturing process, an expert may need to travel to the worksite. There may be numerous technicians available but only a few experts. Augmented reality can reduce this expense and let an expert see the issue through the eyes of a technician. This can let them support and inspect from anywhere in the world. They can also guide the technician about the feature they may be interested in.

5.5.4 Virtual Reality in Industry 4.0

Virtual reality will be playing a leading role in Industry 4.0. It opens various avenues to develop innovative solutions as manufacturing units equip themselves with smart machines, high connectivity, data intelligence platforms, and simulation tools. These changes will help in refining production capabilities and meeting customer requirements [5]

5.6 USES OF VIRTUAL REALITY APPLICATIONS IN INDUSTRY 4.0

5.6.1 Training

Virtual reality helps the organisation in providing its employees with real surroundings virtually. These drills are extremely safe. Training to eliminate the distractions like humans, noises, and other obstacles helps the workers to focus on their work and increase productivity. This training can also help them to deal with real-time difficulties.

5.6.2 Factory Planning

While building a new plant or revamping the current plant, immense efforts are needed in designing, testing, and then in trials. Virtual plants can help in these scenarios. They can be tested as many times as desired so that the flaws in the entire system can be pointed out and corrected. The entire plant can be designed from scratch, and changes can be made as desired.

5.6.3 Inspection

Safety and routine inspections can be carried out by trained experts through a virtual manufacturing process environment. This needs to be done, as manual inspections may miss crucial checking. With virtual reality, it becomes easy for experts to take minute details into consideration.

Virtual reality systems use computer modelling and simulation technologies to create virtual environments, allowing users to interact with them and have an immersive experience. This can be an effective way for people to learn about the world. VR expands human perception by combining multimedia, sensors, displays, human-machine interaction, ergonomics, simulation, computer graphics, and artificial intelligence technologies. Education, healthcare, entertainment, culture, sports, engineering, the military, and other industries have all embraced virtual reality. Although virtual reality has been around in some form for more than half a century, it has only lately evolved into a useful tool for the industrial field. As a result, virtual reality–based manufacturing has gotten a lot of interest as a new technology. Modern manufacturing businesses are under a lot of pressure from global competition because of the rapid advancement of production technology and the frequent changes in client needs. Virtual manufacturing is emerging as a critical technology for dealing with competitive pressures and allowing enterprise transformation and upgrade. Virtual manufacturing has the potential to be useful at any stage of the product lifecycle. Virtual manufacturing, for example, can examine future manufacturing processes and other activities in the product lifecycle during the product design phase to guarantee that product design quality is optimised and production efficiency is maximised [6].

5.7 AR AND VR USED FOR CHANGES IN MANUFACTURING SYSTEMS

AR/VR appears to have several characteristics that make it suited for supporting manufacturing innovations.

Collaboration: Augmented reality is utilised to bring the real and virtual worlds together. As a result, it can be used to visualise future modifications by overlaying them on the current production system. Virtual reality is used for group interactions as well as cooperation within the virtual environment. As a result, it appears to be appropriate for planning, simulating, and evaluating potential changes before implementing them.

Immersion: AR/VR creates the immersion of being physically present in a non-physical or only partially physical world, making it simpler to comprehend modifications, the requirements of acceptance, the realisation process, and the consequences of changes in manufacturing systems.

Annotation: Objects of interest can be highlighted virtually and annotations can be attached to them in order to support subsequent planning steps. In the case of augmented reality, the data can be directly linked to the machines in the real production system. This enables the creation of a knowledge base based on actual circumstances. Annotations can also be added in VR; however, they are based on the underlying 3D model as well as the users' experience.

Interactions: Engineering changes are frequently based on interactions between the involved participants. Misconceptions can be reduced by visualising the object of consideration (for example, the production system area). Interactions using AR may be more difficult due to the higher cost of connecting, for example, numerous smart glasses to share a single view. Interactions with VR are easier, for example, CAVE (Cave Automatic Virtual Environment) is a virtual reality (VR) environment that consists of a cube-shaped VR room or a room-scale area with projection screens on the walls, floors, and ceilings. The user may wear a VR headset or heads-up display and interact with input devices such as wands, joysticks, or data gloves [7].

Furthermore, current techniques have not investigated the possibilities of specific AR/VR technologies in relation to specific industrial or planning procedures or phases. For example, one technology, such as augmented reality on tablets, may be appropriate for preliminary system analysis, whereas VR-based virtual environments may be appropriate for further planning stages. As a result, it is necessary to assess the possibility of linking these technologies with respect to different phases.

Augmented reality and virtual reality technologies increase flexibility by providing the essential information to complete the work task and train operators, resulting in reduction of errors and shorter task times.

A human-computer interaction tool that overlays digital information in the real-world environment in real time is known as augmented reality. This technology can benefit the operator by providing the relevant data to the worker, effectively turning it into a digital assistance system to eliminate human errors. AR systems have a wide range of applications; they can help with things like picking items in a warehouse or providing maintenance instructions via mobile devices.

Virtual reality is a human-computer interface that allows users to engage with virtual environments by replicating them using a real-time computational interface and numerous sensory channels. In industry, VR technology can support understanding of processes and the calculation of parameters and variables based on simulation. Part models can be transformed into interactive virtual simulations to train operators in difficult assembly jobs during the product assembly stage. It also enables the visualisation and virtual usage of elements that are out of reach of the user, as well as the safe operation of hazardous equipment [8].

5.8   AR/VR APPLICATIONS IN OTHER MANUFACTURING AREAS

The effect of AR/VR is truly revolutionary and is creating virtual simulation in all aspects of manufacturing areas. AR/VR is creating positive changes for a vast range of manufacturing processes, from design and prototyping to final production and assembly.

5.8.1   DESIGN AND PROTOTYPING

VR can actually demonstrate how a product would look without creating a physical prototype. Virtual prototyping allows for cheaper and faster design and modification of parts or elements and full models of the products which are to be constructed later for various industries, including automotive, aerospace and aviation, healthcare, and electronics. It is mainly used for visualising, testing, and modifying three-dimensional prototypes in the virtual environment.

Ford Motors has been doing virtual design for a long time and is building fewer physical models and making better choices earlier in the design process. In the old days, cars were designed with pencil and paper, and in modern times, designers relied on computer-aided design tools. In the later case, the design of a three-dimensional car was done on a two-dimensional plane only. With the help of VR, we can now sketch out our ideas in three dimensions with a digital wand. The Ford Immersive Vehicle Environment (FIVE) is a system that employs an 80-inch, 4K monitor connected to a computer, where the viewer wears a VR headset to explore the vehicle with 3D visualisation software. State-of-the-art motion sensors scattered around the room map the movements of users in relation to the virtual model of the car they are designing. We can hold a real steering wheel and then add other key components and switches in the physical world, then match the physical world to the virtual world. That way, when you reach for a switch, you feel it physically and see it virtually as a beautiful, properly rendered virtual vehicle. You can truly sense the proportion and design theme and feel the proper placement of interior controls in a vehicle. This means that teams at Ford can combine engineering practice with the voice of the customer by allowing experts to see things from their perspective. FIVE enables the perception of reality to be altered, and this enables us to experience the car from the standpoint of a taller man or a shorter woman.

5.8.2   INVENTORY MANAGEMENT

A majority of warehouses still use paper-based management systems, but such an approach is slow and inaccurate. A large number of high-technology systems for optimising warehouse operations are available today. These systems help with the labour in performing tasks effectively and quickly result in quality and productivity. Radio frequency picking and artificial intelligence can streamline the complex process of managing inventory. But the task of picking up the product from the warehouse still involves manual labour.

AR technology helps eliminate confusion and makes this process quick and precise. A warehouse worker holding an iPad or wearing a Microsoft HoloLens (or any other headset, for that matter) gets instructions about the exact location of a particular item and is guided to the very aisle and shelf where it is stored. No more guesswork and getting lost amidst the similar-looking shelves—anyone who has ever been inside an industrial warehouse can understand the value of this solution. The augmented reality app captures an image of the environment with the camera in a tablet, smart phone, head-mounted display, or AR smart glasses. The captured image is scanned to identify where to place additional information. The AR software requests specific content to show to the user. The augmented reality application assembles an integral image of the physical environment and overlaid additional information. Vision picking is advantageous because all the operations can be performed by using AR smart glasses without using additional input devices.

The AR application gives the user tips in the form of images or text so that they can quickly select the required item in the right amount. After the selection of the correct item, the picker confirms this action. And already when the correct object is selected, the program shows which bin it should be put in. The whole process of vision picking is built in such a way that it is almost impossible for the user to make a mistake. At each stage, the operator can confirm actions either by reading barcodes with AR smart glasses or by using a voice confirmation system. At the same time, the operator can continue collecting orders and is not distracted by any actions. The entire process of collecting and confirming orders in the vision picking system is as smooth and intuitive as possible.

DHL launched a project of using augmented reality in the warehouse to collect orders. The employees of the warehouse wore AR smart glasses, such as Vuzix and Google Glass. These AR smart glasses show the optimal route for collecting orders and information about the items and their quantity and also indicate in which bin to put the picked items. The barcode scanning feature of the augmented reality app allows pickers to make sure that they choose the right items. Thus, the operator smoothly and quickly picks the required items. Evaluation of the AR project results showed an increase in the efficiency of the picking process by 25%. The interfaces of the AR applications are intuitive, so users easily understand what to do and almost don't need training. Thus, the use of augmented reality allowed DHL to reduce the time for employee training by 50%. Such performance proves the effectiveness of AR application in large warehouse management systems. Accordingly, after a pilot testing of this augmented reality platform, DHL has launched a global implementation of AR vision picking technology in their warehouse processes.

5.8.3 Prevention of Accidents and Disruptions

VR is helping to predict and evade the hazards and disruption risks associated with the use of an assembly line. By simulating the production environment, manufacturing companies can indicate potential threats and eliminate them long before they even arise. The value of this solution is difficult to overlook, since it helps reduce downtime as well as repair and maintenance expenses and enhances employee security.

Using VR applications for training is a great application of virtual reality in manufacturing. VR training is being used in many industries, but manufacturing especially benefits from virtual reality development. Training for a highly hazardous job has rather steep requirements as to the safety of the trainees. Such training is often hard to organise, as, on the one hand, the trainees have no practical skills, which make them more exposed to danger. On the other hand, the use of simulators has limited possibilities and ultimately can hardly give the feel of the actual work. In this case, virtual reality again proves its outstanding ability to place the trainees in a realistic environment while reducing the risk of injury to zero.

Gabler Engineering uses virtual reality in many areas of its operation. Particularly, the company extensively uses VR training for operators of the machines that it produces. Assisted by VR technology, the company inspects production lines for potential hazards and thus helps ensure safety and quality. The company estimated that training that usually takes a week and involves six people to assist the trainee. In virtual reality, it can be completed in 20 minutes. In addition, there is no need to assign additional personnel to the VR training, as no risk is involved.

5.8.4 Inspection and Maintenance

Today, machines are becoming too complex to be inspected by regular factory personnel. Often, for any routine or emergency inspection, companies need to invite experts from the plant that produced the equipment. Such inspection visits usually involve high costs and careful advance planning and scheduling.

Inspectors' time is often booked out for weeks, and it may be difficult to invite them when a machine suddenly malfunctions. Also, the travel and accommodation costs are often to be borne by the inviting side. With virtual reality, the factory does not need to bring the inspector to the machine. Instead, it can bring the machine to the inspector, and in no time, too. By creating a VR video of the equipment, the factory can have it inspected remotely with no need of inviting the expert. Imagine the savings both for the inspectors and manufacturers. When the time and costs of travel are removed from the calculation, inspections and remote maintenance can be done in virtual reality much faster, reducing interruptions in the production cycle to the minimum.

5.9 AR/VR APPLICATIONS IN VARIOUS AREAS

The number of applications utilising augmented reality is increasing continuously, and the benefits can be seen clearly in various fields, including healthcare, business, education, and amusement [9].

5.9.1 HEALTHCARE

The potential uses for these technologies in healthcare are obvious, and we can expect to see many of these use cases transition from trials and pilots and gradually into general use. Virtual reality has already been adopted in therapy, where it is used to treat patients with phobias and anxiety disorders. Combined with biosensors that monitor physiological reactions like heart rate and perspiration, therapists can get a better understanding of how patients react to stressful situations in a safe, virtual environment. VR is also used to help people with autism develop social and communication skills, as well as to diagnose patients with visual or cognitive impairments by tracking their eye movement.

The adoption of AR in healthcare is forecast to grow even more quickly. AR can be used by surgeons—both in the theatre and in training—to alert them to risks or hazards while they are working. One app that has been developed uses AR to guide users towards defibrillator devices, should they need one when they are out in public. Another one helps nurses to find patients' veins and avoid accidentally sticking needles where they aren't wanted. As these innovations and others like them lead to improved patient outcomes and reduced cost of treatment, they are likely to become increasingly widespread.

Augmented reality and virtual reality have revolutionised the healthcare field. Accuvein, an AR device that is a handheld scanner that projects an image of the skin of the valves, veins, and bifurcations underneath, helps healthcare professionals find a vein while giving an injection. Another medical AR application called Eyedecide uses camera display for simulating the impact of specific conditions on a person's vision. With the help of AR, now pharmaceutical companies can provide more useful and innovative drug information. Instead of reading long descriptions of bottles, patients can now see the use and actions of drugs through 3D in front of their eyes [10].

5.9.2 EDUCATION

Augmented and virtual reality have the potential to bring more students into the classroom and create more engaging and exciting classroom experiences. Using VR, students in healthcare facilities or underdeveloped rural areas could participate in the classroom virtually in a much more full and rich way than in traditional online classes. Using AR and VR, teachers could create an immersive learning environment, allowing students to explore in many areas or unfamiliar parts of the globe. Many schools and universities started funding these types of technologies to make learning more inclusive and engaging.

Augmented and virtual reality have made education more interactive and fun filled. Many AR apps are being developed that embed text, images, and videos as well as real-world curricula. Google has announced Expeditions, a virtual reality platform for classrooms. Students can use Cardboard to take guided tours to many cities and also inaccessible places like space.

In terms of higher education, AR apps and 3D glasses are very useful. Colouring books by Disney have made colouring activities very interesting, as students can see their coloured objects come alive on pages by scanning them. Western University of Pomona, California, recently opened a virtual reality centre for medical students. It has various tools, including a digital dissection table. A company called Touch Surgery also uses augmented reality technology to train surgeons in various complicated surgeries.

5.9.3 CULTURE AND TOURISM

By providing additional imagery and historical or cultural details, AR has the ability to transform visitor experiences with city landmarks. Many countries have installed augmented reality screens throughout their historic places in order to illustrate what the places looked like in the Middle Ages, and other cities could pursue similar technologies to highlight their historic pasts.

Both AR and VR have enhanced the traveller's experience and changed the way they see the world. Now people can plan trips seamlessly with the help of AR technology. With the increase in the number of smartphone users, the traditional method of planning a trip is slowly becoming outdated. Now at the tip of our finger, we can get all the information required for the particular place that we are going to visit so we can plan accordingly. We can go through a virtual tour of the hotel room that we want to book or see the surrounding visiting areas of the particular place. Google's AR app, World Lens, has made it easy for users to aim their smart phones at signboards and automatically translate them. This application is very useful for international business travellers. In India, the governments of Kerala and Gujarat have already implemented augmented reality in tourism [11].

5.9.4 GAMING

Gaming is perhaps the only industry that has utilised the potential of AR at its best. AR has actually changed the way the gamers have played games to date, and with the release of Pokémon Go, users have the experience of playing an augmented reality game. We can step out of our house with our device and play the game and see our favourite gaming characters coming alive on the screen. The best part of AR games is that they make us active instead of lazy. We have to move around and play the game. Users get an immersive experience of games with AR technology that they have never had before.

5.9.5 REAL ESTATE

The real estate sector is flourishing day by day, and realtors are looking for new technologies through which they can provide an immersive experience to their prospective clients and customers. Augmented reality is a great tool that is helping both realtors and customers. Through AR, buyers can just aim their smartphones at the property on sale and get all the relevant information about it. They can see the position of their property and the look and feel of it after it is completed. The buyers need not enter the house. They can just pass by the construction site and use their smartphones to see the overlaid data that is visible to them. This actually helps them to make the right purchase decision. Many AR applications have been developed by various realtors for their use and also for their customers' benefit.

5.9.6 EMERGENCY MANAGEMENT

AR can improve responders' knowledge of their surroundings in order to rescue residents in need, much as the rise of data visualisation has enhanced situational awareness during emergency situations. Residents could plot their locations on an interactive map distributed to responders, which would then show workers the safest rescue routes and identify particularly hazardous areas.

5.9.7 MENTAL HEALTH SERVICES

Recent research has indicated that AR and VR have immense potential for treating mental health problems like anxiety and stress disorders, creating opportunities for users to confront their fears in exposure therapies. And VR simulations can also help residents with stress disorder practice job interviews, learning to manage their symptoms in high-stress environments. By furnishing social workers with these tools, cities can improve their mental health treatment [12].

5.10 AUGMENTED AND VIRTUAL REALITY DURING COVID

During the pandemic, many industrial sectors and emergency management teams used AR/VR technology in their respective fields for safe operation.

Many manufacturers look to implement AR/VR technology to address the labour issues created by COVID-19; operations managers should pay close attention to the technology infrastructure surrounding any AR/VR platforms they assess for use in their plants. Manufacturers are overcoming their growth limitations by leveraging cloud-based (or remote server–based) AR/VR platforms powered with distributed cloud architectures and 3D vision-based artificial intelligence. These cloud platforms provide the desired performance and scalability to drive innovation in the industry at speed and scale [13].

The technology allows automotive designers and manufacturers to conduct real-time 3D visualisation and CAD for design and manufacturing, run faster training cycles, and enable professionals to work at drastically higher levels. In fact, some manufacturers report minimised errors using AR/VR through instructions overlay, remote assistance, better planning, and visualisation. This has resulted in productivity increases in some instances.

REFERENCES

[1] R. Ben Khalifa, K. Tliba, M. L. Thierno Diallo, O. Penas, N. Ben Yahia, and J. Y. Choley, "Modeling and management of human resources in the reconfiguration of production system in Industry 4.0 by neural networks," *2019 Int. Conf. Signal, Control Commun. SCC 2019*, pp. 246–249, 2019, doi:10.1109/SCC47175.2019.9116104.

[2] L. Damiani, M. Demartini, G. Guizzi, R. Revetria, and F. Tonelli, "Augmented and virtual reality applications in industrial systems: A qualitative review towards the Industry 4.0 era," *IFAC-PapersOnLine*, vol. 51, no. 11, pp. 624–630, 2018, doi:10.1016/j.ifacol.2018.08.388.

[3] A. Mbiriki, C. Katar, and A. Badreddine, "Improvement of security system level in the cyber-physical systems (CPS) architecture," *Proc. Int. Conf. Microelectron. ICM*, vol. 2018-December, no. Icm, pp. 40–43, 2018, doi:10.1109/ICM.2018.8704100.

[4] D. P. Valentina, D. S. Valentina, M. Salvatore, and R. Stefano, "Smart operators: How Industry 4.0 is affecting the worker's performance in manufacturing contexts," *Procedia Comput. Sci.*, vol. 180, no. 2019, pp. 958–967, 2021, doi:10.1016/j.procs.2021.01.347.

[5] D. Antonelli *et al.*, "Tiphys: An open networked platform for higher education on Industry 4.0," *Procedia CIRP*, vol. 79, pp. 706–711, 2019, doi:10.1016/j.procir.2019.02.128.

[6] W. Zhu, X. Fan, and Y. Zhang, "Applications and research trends of digital human models in the manufacturing industry," *Virtual Real. Intell. Hardw.*, vol. 1, no. 6, pp. 558–579, 2019, doi:10.1016/j.vrih.2019.09.005.

[7] C. Siedler, M. Glatt, P. Weber, A. Ebert, and J. C. Aurich, "Engineering changes in manufacturing systems supported by AR/VR collaboration," *Procedia CIRP*, vol. 96, pp. 307–312, 2020, doi:10.1016/j.procir.2021.01.092.

[8] D. V. Enrique, J. C. M. Druczkoski, T. M. Lima, and F. Charrua-Santos, "Advantages and difficulties of implementing Industry 4.0 technologies for labor flexibility," *Procedia Comput. Sci.*, vol. 181, pp. 347–352, 2021, doi:10.1016/j.procs.2021.01.177.

[9] A. O. Alkhamisi and M. M. Monowar, "Rise of augmented reality: Current and future application areas," *Int. J. Internet Distrib. Syst.*, vol. 01, no. 04, pp. 25–34, 2013, doi:10.4236/ijids.2013.14005.

[10] S. K. Ong, M. L. Yuan, and A. Y. C. Nee, "Augmented reality applications in manufacturing: A survey," *Int. J. Prod. Res.*, vol. 46, no. 10, pp. 2707–2742, 2008, doi:10.1080/00207540601064773.

[11] A. N. Sküng, "A brief introduction of VR and AR applications in manufacturing," in *Virtual and Augmented Reality Applications in Manufacturing*, Springer-Verlag, London, 2004.

[12] A. Y. C. Nee, S. K. Ong, G. Chryssolouris, and D. Mourtzis, "Augmented reality applications in design and manufacturing," *CIRP Ann.—Manuf. Technol.*, vol. 61, no. 2, pp. 657–679, 2012, doi:10.1016/j.cirp.2012.05.010.

[13] I. Zolotová, P. Papcun, E. Kajáti, M. Miškuf, and J. Mocnej, "Smart and cognitive solutions for Operator 4.0: Laboratory H-CPPS case studies," *Comput. Ind. Eng.*, vol. 139, 2020, doi:10.1016/j.cie.2018.10.032.

6 Cloud-Based Manufacturing Service Selection Using Simulation Approaches

Vaibhav S. Narwane, Irfan Siddavatam, and Rakesh D. Raut

CONTENTS

6.1 Introduction 49
6.2 Review of Critical Factors in Selection, Development of Framework,
Modelling, and Simulations 50
6.3 Methodology 50
    6.3.1 Condition 1 51
    6.3.2 Condition 2 52
6.4 Results and Discussion 53
6.5 Conclusion and Future Scope 54
References 55

6.1 INTRODUCTION

Globalization truly has had a deep impact on the manufacturing sector. Manufacturing systems have moved from production oriented to service oriented (Xu, 2012) and from mass production to mass customization (Mourtzis et al., 2013). Cloud technology applications in manufacturing have been gaining momentum over the years. Bo Hu Li coined the term "cloud manufacturing" (CM) in 2009. Many researchers have defined CM in their own way, and there is no international standard definition for CM. For example, Wu et al. (2013) define CM as a customer-centric manufacturing model that exploits on-demand access to a shared collection of diversified and distributed manufacturing resources to form temporary, reconfigurable production lines which enhance efficiency, reduce product lifecycle costs, and allow for optimal resource loading in response to variable-demand customer-generated tasking.

CM is being adapted by small and medium-sized companies (SMEs), startup companies, technical firms, heavy-duty machine toolers, the polymer material industry, original equipment manufacturers (OEMs), and so on. According to Haug et al. (2016), large firms are early adopters of CM, but small firms catch up quickly. The benefits of CM are: SMEs can reduce initial investment cost (Marston et al., 2011), it can enhance business portfolios for startup companies (Repschlaeger et al., 2013), it can handle unregulated market sectors in services (Haug et al., 2016), and it has relative advantages and compatibility (Gangwar et al., 2015).

Most of the work on CM is done in selected SMEs, where resource and knowledge sharing become crucial (Adamson et al., 2017). This same point is the motivation for this chapter.

The rest of this chapter is organised as follows: Section 6.2 presents literature on CM. Section 6.3 describes a proposed simulation of system process, followed by results and discussion in Section 6.4. Conclusions and future directions are presented in Section 6.5.

DOI: 10.1201/9781003186670-6

6.2　REVIEW OF CRITICAL FACTORS IN SELECTION, DEVELOPMENT OF FRAMEWORK, MODELLING, AND SIMULATIONS

Cloud manufacturing is adopting and extending cloud computing concepts for manufacturing. The potential impacts of CM are not only for manufacturing but also design and service. Wu et al. (2013) listed the short-term impacts of CM as improved customer need elicitation; less cost; reduced time to market; improved efficiency, service quality, and resources; and ubiquitous access to design information. In the long term, CM will help in collaborative design, distributed manufacturing, and customer creation (Wu et al., 2013). According to Chen and Tsai (2017), CM is going to boost ubiquitous manufacturing (UM) for widespread deployment of manufacturing facilities. Information and computation technology (ICT) systems are supporting global manufacturing and its supply chain; CM can improve its agility and security (Radke and Tseng, 2015). Zhong et al. (2017) listed six critical characteristics of CM: the Internet of things (IoT) of manufacturing, virtual manufacturing, service-oriented manufacturing, efficient collaboration, knowledge-intensive manufacturing, and future social manufacturing. The latest trends in the IoT, cyber-physical systems (CPSs), big data, and artificial intelligence can help improve CM (Tao et al., 2014). The CM model provides big data storage and transmission as well as a processing platform for designers, manufacturers, and users (Liu et al., 2016). CM makes it easier for small and large manufacturing enterprises to scale their production and business according to client demand (Marston et al., 2011). According to Tao et al. (2011), CM improves efficiency, interoperability, sustainability, and flexibility (Wang and Xu, 2013).

6.3　METHODOLOGY

The core idea behind a cloud service selection system is to help uses select the manufacturer or service provider that is the most suitable for their requirements. Figure 6.1 shows the basic framework of the proposed methodology.

The next two subsections explain service and manufacturer selection. CloudAnalyst software supports virtual modelling and simulation of large-scale applications used for simulation.

In real-life scenarios, there are thousands of manufacturers present in the market, and tapping into information on individuals and simultaneously processing that data is an enormous task. Also, doing it manually is practically not possible. Hence cloud technology should be incorporated in such a case. From an application point of view, a selection system can be considered a large-scale application. Cloud technology proves useful in such a case because it presents non-uniform usage patterns. With the help of CloudAnalyst, information regarding the response time of a request, processing time, and cost analysis matrices can be generated.

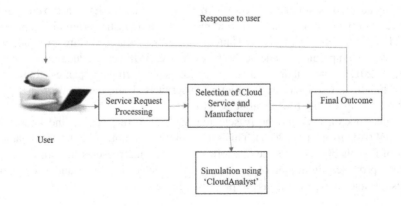

FIGURE 6.1　Basic framework of proposed cloud service selection system.

(*Source*: Author)

For this study, the model proposed is a selection system on the behaviour of Amazon.com, an e-commerce and cloud computing company. As of December 2016, Amazon had over 180 million unique monthly visitors worldwide. It attracted around 35% of its visitors from North America, while Europe contributed 31% and Asia Pacific 24% (source: www.webintravel.com).

We defined four user bases that represent the four regions of the world with described parameters (refer to Figure 6.2). User bases 1, 2, 3, and 4 are the North America, Europe, Asia, and Oceania regions, respectively.

For this simulation, a similar hypothetical application is chosen at 1/10th of the scale of Amazon (source: https://aws.amazon.com). Assumptions made were as follows: users were from only one time zone, and they were more active in the morning hours. Costing was done based on the actual pricing policies of Amazon Cloud services.

Here assessment was done for the nature of data centre to be developed, that is, whether it should be a centralized or decentralized network. Hence two conditions were simulated, first a centralized single data centre and second a decentralized data centre network.

6.3.1 CONDITION 1

In this condition, a single centralized data centre was developed. For simplicity purposes, its configuration was kept on the simpler side, and five single physical hardware units were assigned. Details of this data centre are given in Figures 6.3 and 6.4.

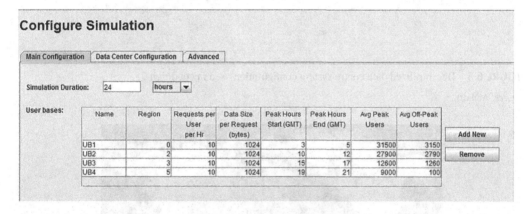

FIGURE 6.2 User bases and their specifications used in experiments.

(*Source*: Author)

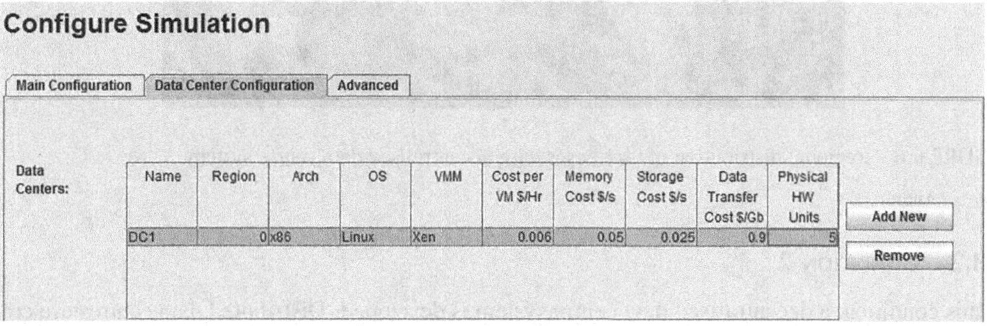

FIGURE 6.3 Data centre configuration used in experiments for condition 1.

(*Source*: Author)

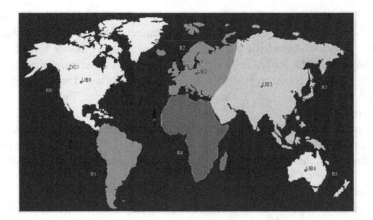

FIGURE 6.4 Regional distributions of user bases with centralised data centre.

(*Source*: Author)

Main Configuration	Data Center Configuration	Advanced

Data Centers:	Name	Region	Arch	OS	VMM	Cost per VM $/Hr	Memory Cost $/s	Storage Cost $/s	Data Transfer Cost $/Gb	Physical HW Units	
	DC1	0	x86	Linux	Xen	0.006	0.05	0.025	0.09	5	Add New
	DC2	2	x86	Linux	Xen	0.007	0.05	0.024	0.09	5	Remove
	DC3	3	x86	Linux	Xen	0.007	0.05	0.025	0.011	5	
	DC4	5	x86	Linux	Xen	0.008	0.05	0.025	0.14	5	

FIGURE 6.5 Decentralized data centre system configuration used in condition 2.

(*Source*: Author)

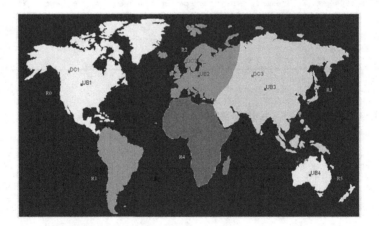

FIGURE 6.6 Regional distribution of user bases with decentralised data centre system.

(*Source*: Author)

6.3.2 CONDITION 2

In this condition, a decentralised data centre system is developed. Distributed data centres are created based on user base regions. For simplicity purposes, its configuration was kept on the simpler side, and five physical hardware units were assigned to each data centre. Details of this data centre are given in Figures 6.5 and 6.6.

6.4 RESULTS AND DISCUSSION

The results of the cloud simulation analysis are shown in Tables 6.1 to 6.5.

These results are the overall values obtained after simulation. Response times for individual user bases and data centre service are given in Table 6.3.

Similarly, in Table 6.4, response times for individual user bases and data centre service times for a decentralised data centre system are given.

TABLE 6.1
Results for Overall Response Time of User Request

Condition	Service Selection Policy	Overall Response Time (ms)		
		Average	Minimum	Maximum
Centralized data centre	Closest data centre	237.66	58.48	510.46
system	Optimum response time	234.22	55.01	562.96
Decentralised data	Closest data centre	62.05	46.90	72.26
centre system	Optimum response time	63.01	52.94	66.66

TABLE 6.2
Results for Data Centre Processing Time of User Request

Condition	Service Selection Policy	Data Centre Processing Time (ms)		
		Average	Minimum	Maximum
Centralized data centre	Closest data centre	6.52	2.64	12.65
system	Optimum response time	6.35	2.57	12.15
Decentralised data	Closest data centre	8.18	2.95	12.90
centre system	Optimum response time	8.12	2.31	12.50

TABLE 6.3
Response Times for Individual User Bases and Data Centre Service Times for Centralised Data Centre System

Service Selection Policy	Parameter		Time (ms)		
			Average	Minimum	Maximum
Closest data centre	User base	UB1	59.05	58.48	59.61
	response time	UB2	352.88	325.52	380.23
		UB3	497.96	485.45	510.46
		UB4	200.40	200.05	200.74
	Data centre service times	DC1	6.52	2.64	12.65
Optimum response time	User base	UB1	59.88	55.01	64.75
	response time	UB2	327.23	318.54	335.91
		UB3	526.66	490.36	562.96
		UB4	216.89	209.76	224.02
	Data centre service times	DC1	6.35	2.57	12.15

TABLE 6.4

Response Times for Individual User Base and Data Centre Service Times for Decentralised Data Centre System

Service Selection Policy	Parameter		Time (ms)		
			Average	Minimum	Maximum
Closest data centre	User base	UB1	60.12	56.28	63.95
	response time	UB2	68.34	64.42	72.26
		UB3	55.45	54.90	55.99
		UB4	51.17	46.90	55.44
	Data centre	DC1	7.00	2.95	11.06
	service times	DC2	11.44	9.98	12.90
		DC3	5.20	4.23	6.16
		DC4	3.12	3.10	3.14
Optimum response time	User base	UB1	66.25	66.15	66.35
	response time	UB2	61.65	56.65	66.66
		UB3	57.49	53.62	61.36
		UB4	55.68	52.94	58.42
	Data centre	DC1	7.04	2.31	11.77
	service times	DC2	11.33	10.17	12.50
		DC3	4.96	3.84	6.07
		DC4	3.00	2.88	3.11

TABLE 6.5

Overall Costing Results Obtained from Simulation of Cloud System

Cost ($)	Data Centre System Structure	
	Centralized	Decentralized
Total virtual machine cost	0.01	0.03
Total data transfer cost	12.26	1.05
Grand total	12.27	1.08

Results obtained for costing are shown in Table 6.5.

By taking a closer look at Tables 6.1 to 6.5, one thing is absolutely clear: the decentralised data centre system structure will be more suitable for the selected system application. Though its data centre processing time is little higher, the response time for user request input and overall costing for the cloud system are much lower.

6.5 CONCLUSION AND FUTURE SCOPE

Cloud-based manufacturing has come a long way since its inception. However, significant challenges of CM include: i) *data management*: It is important to store and analyse big data from all the connected devices for processing and to show real-time results. ii) *Privacy and security*: One of the biggest challenges in CM is to balance privacy concerns and personal data control to provide better services because CM manages large amounts of data, and sometimes it can be sensitive. CM requires privacy policies in order to deal with privacy issues. iii) *Data heterogeneity*: Data heterogeneity is a significant issue that can affect communication performance and the design of communication protocols. iv) *Lack of knowledge about resource pooling*: Companies have less awareness about CM, particularly SMEs. Resource pool creation is an important aspect of CM, and CM simulation tool software is rarely used by researchers.

This chapter will help manufacturers and researchers regarding resource pooling of CM. Simulation analysis of service systems is done using the CloudAnalyst software to gain insight into practical implementation. Practical application of the proposed selection system using cloud technology needs identification of possible hubs of user bases as well as SMEs. Once these hubs are identified, time and cost can be saved significantly. For example, currently China is a manufacturing hub while India is in the process of becoming one; thus, the number of cloud data centres can be increased in this region for better performances. Future work includes real-life applications based on a proposed framework of selection systems and studying environmental effects and energy consumption.

REFERENCES

Adamson, G., Wang, L., Holm, M., & Moore, P. (2017). Cloud manufacturing—A critical review of recent development and future trends. *International Journal of Computer Integrated Manufacturing, 30*(4–5), 347–380.

Chen, T., & Tsai, H. R. (2017). Ubiquitous manufacturing: Current practices, challenges, and opportunities. *Robotics and Computer-Integrated Manufacturing, 45*, 126–132.

Gangwar, H., Date, H., & Ramaswamy, R. (2015). Understanding determinants of cloud computing adoption using an integrated TAM-TOE model. *Journal of Enterprise Information Management, 28*(1), 107–130.

Haug, K. C., Kretschmer, T., & Strobel, T. (2016). Cloud adaptiveness within industry sectors—Measurement and observations. *Telecommunications Policy, 40*(4), 291–306.

Liu, Z., Wang, Y., Cai, L., Cheng, Q., & Zhang, H. (2016). Design and manufacturing model of customized hydrostatic bearing system based on cloud and big data technology. *The International Journal of Advanced Manufacturing Technology, 84*(1–4), 261–273.

Marston, S., Li, Z., Bandyopadhyay, S., Zhang, J., & Ghalsasi, A. (2011). Cloud computing—The business perspective. *Decision support systems, 51*(1), 176–189.

Mourtzis, D., Doukas, M., & Psarommatis, F. (2013). Design and operation of manufacturing networks for mass customisation. *CIRP Annals-Manufacturing Technology, 62*(1), 467–470.

Radke, A. M., & Tseng, M. M. (2015). Design considerations for building distributed supply chain management systems based on cloud computing. *Journal of Manufacturing Science and Engineering, 137*(4), 040906.

Repschlaeger, J., Erek, K., & Zarnekow, R. (2013). Cloud computing adoption: An empirical study of customer preferences among start-up companies. *Electronic Markets, 23*(2), 115–148.

Tao, F., Cheng, Y., Da Xu, L., Zhang, L., & Li, B. H. (2014). CCIoT-CMfg: Cloud computing and internet of things-based cloud manufacturing service system. *IEEE Transactions on Industrial Informatics, 10*(2), 1435–1442.

Tao, F., Zhang, L., Venkatesh, V. C., Luo, Y., & Cheng, Y. (2011). Cloud manufacturing: A computing and service-oriented manufacturing model. *Proceedings of the Institution of Mechanical Engineers, Part B: Journal of Engineering Manufacture, 225*(10), 1969–1976.

Wang, X. V., & Xu, X. W. (2013). An interoperable solution for cloud manufacturing. *Robotics and Computer-Integrated Manufacturing, 29*(4), 232–247.

Wu, D., Greer, M. J., Rosen, D. W., & Schaefer, D. (2013). Cloud manufacturing: Strategic vision and state-of-the-art. *Journal of Manufacturing Systems, 32*(4), 564–579.

Xu, X. (2012). From cloud computing to cloud manufacturing. *Robotics and Computer-integrated Manufacturing, 28*(1), 75–86.

Zhong, R. Y., Xu, X., Klotz, E., & Newman, S. T. (2017). Intelligent manufacturing in the context of industry 4.0: A review. *Engineering, 3*(5), 616–630.

7 Go Unsupervised via Artificial Intelligence

Amit Vishwakarma, G.S. Dangayach,
M.L. Meena, and Sumit Gupta

CONTENTS

7.1 Introduction 57
7.2 Machine Learning 58
7.3 Various Techniques of Machine Learning 58
    7.3.1 Support Vector Machine 58
    7.3.2 *K*-Nearest Neighbour 58
    7.3.3 Decision Tree 58
    7.3.4 Classification and Regression Tree Methodology 58
    7.3.5 Fuzzy Logic 59
7.4 Supervised and Unsupervised Learning 59
    7.4.1 Supervised Learning 59
    7.4.2 Unsupervised Learning 59
    7.4.3 Drug Discovery and Repurposing 60
    7.4.4 Clinical Trials 60
    7.4.5 Epidemic/Pandemic Outbreak Prediction 60
7.5 Semi-Supervised Learning 60
7.6 Reinforcement Learning 61
7.7 Role of Artificial Intelligence via Unsupervised Learning in Healthcare
    Supply Chains 61
    7.7.1 Stakeholder Selection (Customer–Supplier Relation, Vendor Rating) 61
    7.7.2 Risk Mitigation in SCM 61
    7.7.3 Managing Inventory 62
    7.7.4 Production of Customised Product 62
7.8 Conclusion 62
References 62

7.1 INTRODUCTION

The healthcare sector is very dynamic and challenging in nature. It faces various problems and challenges, and pandemics like COVID-19 put additional pressure on it. It is essential to deal with these problems with the latest technology rather than in conventional ways. This includes the application of artificial intelligence (AI) in digital acquisition of data as well as computing infrastructure via machine learning. AI is slowly replacing some healthcare operations and has enormous potential for solving various challenges. It provides ease for doctors during treatment of patients. In critical operations like brain or heart surgery, AI helps in predicting the result and predicting the effects of medicine intake by the patient. AI has potential in drug discovery. The mixing of various chemicals and their effect on the human body is can be estimated, encoding crucial and sufficient information in drug labels. All these operations are internal to the hospital, but AI can also serve the healthcare sector externally.

DOI: 10.1201/9781003186670-7

As the timely delivery of healthcare products is crucial for saving human lives, AI empowers the supply chain via providing transparency and ultimately leads to increased efficiency and effectiveness. Applications of AI cover various aspects of Supply Chain Management (SCM) like the customer–supplier relationship, inventory management, risk analysis, estimation of demand, estimation of production, and estimation of sales (Tirkolaee et al., 2021).

7.2   MACHINE LEARNING

Machine learning is the process in which a machine learns from data that is provided to the machine in various forms, such as images, numerical data, and so on. The machine uses statistics to understand the pattern. Based on its observation, it predicts the outcome. This outcome is beneficial because people can use it to make future decisions.

To predict the outcome, the machine uses algorithms and modelling of data (Breiman, 2001). In addition to this, machine learning has the capability to handle big data. Those data come from previous records and reports and digitally from various electronic instruments.

7.3   VARIOUS TECHNIQUES OF MACHINE LEARNING

This section includes various methods which are generally used in machine learning. It consists of a total of six techniques that are briefly described (Shailaja et al., 2016).

7.3.1   SUPPORT VECTOR MACHINE

This technique mainly deals with a supervised learning model. Analysis of the provided data consists of classification and implementation of regression analysis.

7.3.2   K-NEAREST NEIGHBOUR

This technique mainly provides non-parametric classification of data. The output depends on whether there is a requirement for classification or regression.

7.3.3   DECISION TREE

This technique mainly deals with a supervised learning model. In it, the leaves and branches symbolise class labels and conjunction of features, respectively.

7.3.4   CLASSIFICATION AND REGRESSION TREE METHODOLOGY

Classification and regression tree methodology (CART) is a technique in which the target variable is represented as categorical for classification and continuous for regression trees.

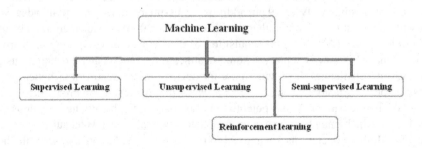

FIGURE 7.1   Types of machine learning.

7.3.5 Fuzzy Logic

This technique is based on fuzzy theory, in which the truth value of a variable lies between zero and one.

The predictive modelling of machine learning is helpful for the healthcare sector. There are many opportunities for machine learning, like predicting the various risks of patients and informing the relevant authorities of the same (Parikh et al., 2016).

7.4 SUPERVISED AND UNSUPERVISED LEARNING

Supervised learning and unsupervised learning are the two basic approaches to AI and machine learning. Both approaches have advantages and disadvantages, but there are some subtle differences between them.

7.4.1 Supervised Learning

In this type of learning, models are trained on labelled data and output is predicted based on the features learned. Structured data are fed to machine learning models, where the machine is able to recognise the pattern between the sample and extracted features. This type of leaning is useful to solve linear and nonlinear issues (Sacco et al., 2016).

Various authors contributed to implement supervised learning in healthcare. In this direction, Salman et al. (2017) implemented a novel method to examine patients with heart disease by assessing and scoring the data collected by various sensors. The authors utilised electronic equipment like electrocardiograms and SpO_2 sensors to examine the desired parameters of the patient's body. Patil et al. (2015) proposed a framework for examining the health of ICU patients. In this work, ZigBee technology was employed. Data acquisition was done by ZigBee. The advantage of this technology is that it reduces data complexity since its power consumption is lower compared to other technology.

7.4.2 Unsupervised Learning

As its name suggests, unsupervised learning does not require human intervention, as the models are trained using unstructured data samples. Machines operate with large data sets with the help of a program to reach a decision (target).

Unsupervised learning is more complex than supervised learning and requires a much bigger data set. However, it is quite useful and has many more applications than supervised

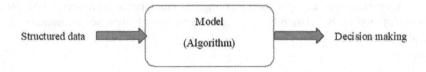

FIGURE 7.2 Supervised learning.

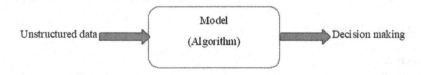

FIGURE 7.3 Unsupervised learning.

learning. It is the process of turning observations in the laboratory, clinic, and community into interventions that improve the health of individuals and the public using diagnostics and therapeutics for medical procedures and behavioural changes. Translation can be used in the following ways

- Drug discovery and repurposing
- Clinical trials
- Epidemic/pandemic outbreak prediction

7.4.3 DRUG DISCOVERY AND REPURPOSING

Hinton et al. (2006) applied unsupervised learning proposed algorithms called deep belief networks, which are useful for computational drug design. Gawehn et al. (2016) discussed unsupervised restricted Boltzmann machine (RBM) pre-training. This method reduces network overfitting by initialising weights utilising unlabelled samples. These properties help in big data analysis for drug discovery.

Unsupervised learning is used in drug design and discovery. In addition to this, it is also used in revealing information on drug labels (Bisgin et al., 2011). This leads to prevention of adverse drug reactions (ADRs) and helps in identifying drug efficacy.

7.4.4 CLINICAL TRIALS

Unsupervised learning enables pattern recognition that is useful in performing clinical trials. Moreover, it is also helpful in predicting the response of any particular drug to be injected into a patient's body. This enables future treatment processes.

The available literature showed that unsupervised learning is applicable in many organ treatments like brain injury, heart attack, tumour identification, and others.

Elmer et al. (2020) used unsupervised learning to identify brain injury phenotypes. This study also tested interactions between phenotype and targeted temperature management (TTM), hemodynamic management, and cardiac catheterisation in models predicting recovery.

7.4.5 EPIDEMIC/PANDEMIC OUTBREAK PREDICTION

The unsupervised learning is used to predict the effect of various diseases in epidemic/pandemic outbreaks. In addition to this, unsupervised learning is also able to predict the peak and duration of epidemics. Some studies are available in the literature in which unsupervised learning predictions were accurate, such as for the morbidity rate of dengue haemorrhagic fever in Thailand (Kesorn et al., 2015) Apart from this, unsupervised learning has much potential in serving people's health, such as in exploration of the relation between genes and possible diseases for humans. Regression models deal with such cases.

7.5   SEMI-SUPERVISED LEARNING

As its name suggests, this type of learning method is used for labelled as well as unlabelled data sets. The model is trained using a few labelled data sets, and the result is predicted for an unlabelled data set.

This technique has two types:

1. Clustering
2. Classification

7.6 REINFORCEMENT LEARNING

In this type of learning, the model uses the concept of reward and punishment (Gentsch, 2018). When the model predicts the correct result, it rewards itself; otherwise, it is punished. Hence the model uses feedback to train itself. The following are algorithms which use the reinforcement concept:

- Temporal difference
- Q-learning
- Deep adversarial network (DAN)

With reinforcement learning, the optimal solution is unknown to the system at the beginning of the learning phase and therefore must be determined iteratively. In this process, sensible approaches are rewarded, and wrong steps tend to be punished. With this approach, it is possible for the system to take into account complex environmental influences and react accordingly. Hence, the system finds its own solutions autonomously through directional rewards and punishments.

7.7 ROLE OF ARTIFICIAL INTELLIGENCE VIA UNSUPERVISED LEARNING IN HEALTHCARE SUPPLY CHAINS

Today AI contributes in various aspects of the supply chain. This section discusses the implementation of AI in various potential areas of SCM.

7.7.1 STAKEHOLDER SELECTION (CUSTOMER–SUPPLIER RELATION, VENDOR RATING)

Mori et al. (2010) discussed application of AI in predicting customer–supplier relationships, as both customer and supplier are important stakeholders in the supply chain. The authors proposed a computational method which found the business profiles of various persons and firms. The authors constructed an algorithm (via machine learning) which predicted future customer–supplier relationships.

Guo et al. (2008) contributed in the area of supplier selection of supply chain management concerning the purchasing element of SCM. The authors used support vector machine technology with the combination of the decision tree method of machine learning for decision making in supply selection. They considered several parameters for deciding priorities like feature selection and multiclass classification.

Muralidharan et al. (2002) proposed a model for supplier rating that consisted of multi-criteria and decisions to be made based on the supplier rating. Various attributes were used, such as supplier evaluation quality, delivery, price, technical capability, financial, past performance attitude, facility, flexibility, and service. Moreover, individual performance of the supplier was also evaluated. Based on the criteria, the performance was evaluated and a decision made. All these are helpful in purchasing material from a suitable vendor.

7.7.2 RISK MITIGATION IN SCM

One important aspect of the supply chain is risk management (Baryannis et al., 2019). Risk is created because of an unexpected event occurring. This includes events like fluctuation of demand, improper planning, and delays in delivery of material. Here AI has the potential for mitigating the risk. Based on the previous data, it develops an algorithm which can partially predict the future, which would be followed by proper decision making. These practices lead to success in mitigating the risk.

7.7.3 Managing Inventory

Inventory control is an important aspect of the supply chain, as it deals with storage of raw material as well as finished product. Inventory control carries a substantial cost for SCM.

Conventionally, the inventory manager is an experienced person in the firm. Based on his or her previous experience, he or she makes a judgement on the quantity to carry in inventory. The past data of the firm are available, so AI can help in estimating the right quantity that is desired in the inventory (Gumus et al. 2010). A multi-echelon inventory management model followed by a neural network of AI simulation is suited for solving this type of problem. This would precisely estimate the demand and lead time for the material. Ultimately it leads to an increase in the performance of SCM and optimal cost incurred in inventory.

7.7.4 Production of Customised Product

This is the era of customised rather than generalised product, and customised product increases the complexity of production. It requires accurate process planning and scheduling operation. Here a neural network using AI has the capability to satisfy customised needs (Chen et al., 2012). This is done by taking the previous data of the firm and managing decision making in the inventory to reduce work in process inventory and bring transparency to the different stages to reduce the complexity in production of a customised product.

It had been found that AI can be used in supplier selection, vendor rating, inventory management, risk mitigation, and production of customised part. Big data available in a firm from previous work by different departments has proved to be an asset for implementing machine learning techniques. These are quite helpful for decision making and reduce human effort, and they can be utilised in other areas to improve the efficiency of the supply chain.

7.8 CONCLUSION

This chapter looked at the role of artificial intelligence in healthcare and the supply chain. Its major focus was unsupervised learning by AI. In healthcare, various machine learning techniques are very useful for doctors as well as for patients during treatment. Machine learning supports treatment of a wide range of diseases, such as detection of heart disease, analysis of diabetic diseases, disclosure of breast cancer, and diagnosis of thyroid disorder. There are certain techniques of machine learning via AI that are contributing in healthcare, for example, support vector machines, K-nearest neighbour, decision trees, CART, and fuzzy logic. All these techniques provide algorithms for predicting or estimating possible outcomes. Moreover, they are also helping in decision making based on previous recorded data. Unsupervised learning has a wide range of applications, and it deals with more complicated task as compared to supervised learning. Its major applications are in drug discovery and performing clinical trials. It has also contributed to general health via prediction of epidemic outbreaks. In addition to this, AI controls many elements of the supply chain, including stakeholder selection, risk mitigation, managing inventory, and production of customised product. All these tasks are efficiently handled by AI as compared to conventional methods. Ultimately, the incorporation of AI increases the overall performance of SCM.

In future work, unsupervised learning can contribute to healthcare by handling more complicated diseases, and it should provide more accurate results that will help doctors make decisions. In contrast, unsupervised leaning should also contribute to other elements of SCM. There is a need to develop more AI algorithms which can explore other uncertainties in SCM.

REFERENCES

Baryannis, G., Validi, S., Dani, S., & Antoniou, G. (2019). Supply chain risk management and artificial intelligence: State of the art and future research directions. *International Journal of Production Research*, 57(7), 2179–2202.

Bisgin, H., Liu, Z., Fang, H., Xu, X., & Tong, W. (2011, December). Mining FDA drug labels using an unsupervised learning technique-topic modeling. In *BMC Bioinformatics* (Vol. 12, No. 10, pp. 1–8). BioMed Central.

Breiman, L. (2001). Statistical modeling: The two cultures (with comments and a rejoinder by the author). *Statistical science*, 16(3), 199–231.

Chen, M.-K., Tai, T.-W., & Hung, T.-Y. (2012). Component selection system for green supply chain. *Expert Systems with Applications*, 39(5), 5687–5701.

Elmer, J., Coppler, P. J., May, T. L., Hirsch, K., Faro, J., Solanki, P., . . . & Callaway, C. W. (2020). Unsupervised learning of early post-arrest brain injury phenotypes. *Resuscitation*, 153, 154–160.

Gawehn, E., Hiss, J. A., & Schneider, G. (2016). Deep learning in drug discovery. *Molecular Informatics*, 35(1), 3–14.

Gentsch, P. 2018. Künstliche Intelligenz für Sales, Marketing und Service. *Mit AI und ots zu einem Algorithmic Business—Konzepte, Technologien und Best Practices* [e-book]. Wiesbaden: SpringerGabler. <https://link.springer.com/content/pdf/10.1007%2F978-3-658-19147-4.pdf>

Gumus, A. T., Guneri, A. F., & Ulengin, F. (2010). A new methodology for multi-echelon inventory management in stochastic and neuro-fuzzy environments. *International Journal of Production Economics*, 128(1), 248–260.

Guo, X., Yuan, Z., & Tian, B. (2008). Supplier selection based on hierarchical potential support vector machine. *Expert Systems with Applications*, 36(3), 6978–6985.

Hinton, G. E., & Salakhutdinov, R. R. (2006). Reducing the dimensionality of data with neural networks. *Science*, 313(5786), 504–507.

Kesorn, K., Ongruk, P., Chompoosri, J., Phumee, A., Thavara, U., Tawatsin, A., & Siriyasatien, P. (2015). Morbidity rate prediction of dengue hemorrhagic fever (DHF) using the support vector machine and the aedes aegypti infection rate in similar climates and geographical areas. *PLoS ONE*, 10, e0125049.

Mori, J., Kajikawa, Y., Sakata, I., & Kashima, H. (2010, December). Predicting customer-supplier relationships using network-based features. In *2010 IEEE International Conference on Industrial Engineering and Engineering Management* (pp. 1916–1920). IEEE.

Muralidharan, C., Anantharaman, N., & Deshmukh, S. G. (2002). A multi-criteria group decision-making model for supplier rating. *Journal of Supply Chain Management*, 38(3), 22–33.

Parikh, A. P., Täckström, O., Das, D., & Uszkoreit, J. (2016). A decomposable attention model for natural language inference. *arXiv preprint arXiv:1606.01933*.

Patil, H. V., & Umale, V. M. (2015). Arduino based wireless biomedical parameter monitoring system using Zigbee. *International Journal of Engineering Trends and Technology (IJETT)*, 28(1).

Sacco, R. L., Roth, G. A., Reddy, K. S., Arnett, D. K., Bonita, R., Gaziano, T. A., et al. (2016). The heart of 25 by 25: Achieving the goal of reducing global and regional premature deaths from cardiovascular diseases and stroke: A modeling study from the American Heart Association and World Heart Federation. *Circulation*, 133(23), e674–e690.

Salman, O. H., Zaidan, A. A., Zaidan, B. B., & Hashim, M. (2017). Novel methodology for triage and prioritizing using "big data" patients with chronic heart diseases through telemedicine environmental. *International Journal of Information Technology & Decision Making*, 16(5), 1211–1245.

Shailaja, K., & Anuradha, B. (2016, December). Effective face recognition using deep learning based linear discriminant classification. In *2016 IEEE International Conference on Computational Intelligence and Computing Research (ICCIC)* (pp. 1–6). IEEE.

Tirkolaee, E. B., Sadeghi, S., Mooseloo, F. M., Vandchali, H. R., & Aeini, S. (2021). Application of machine learning in supply chain management: A comprehensive overview of the main areas. *Mathematical Problems in Engineering*, 2021.

8 Integration of Cyber-Physical Systems for Flexible Systems

Thirupathi Samala, Vijaya Kumar Manupati,
Bethalam Brahma Sai Nikhilesh, and Jose Machado

CONTENTS

8.1 Introduction 65
8.2 Literature Review 66
8.3 Proposed Framework for Integrated Cyber-Physical Systems 67
8.4 Enablers 68
    8.4.1 Manufacturing Data Collection from Sensors/Actuators 68
    8.4.2 IoT/IIoT 68
8.5 Production Planning and Control 68
    8.5.1 Advanced Analytics 68
    8.5.2 Artificial Intelligence, Machine Learning, and Deep Learning 69
    8.5.3 Modeling, Simulation, and Optimization 69
8.6 Maintenance Management 69
    8.6.1 Observe/Analyze 69
    8.6.2 Descriptive Analytics 70
    8.6.3 Predictive Analytics 70
    8.6.4 Prescriptive Analytics 70
8.7 Flexible Configurations 70
8.8 Managerial Implications 71
8.9 Conclusions 71
References 72

8.1 INTRODUCTION

With the advancement of sensors, actuators, data acquisition systems, communication, and the latest network technologies, the manufacturing field is transforming in the digital age. Hence, there is a need to integrate cyber physical systems (CPSs) with traditional production planning and control (PPC) and maintenance management (MM) for manufacturing industries. CPS is the integration of physical processes with computation, information, and communication technologies, as the systems are immersed with the physical components and interact with those physical processes. Generally, the physical part consists of human/material/machine/environment, which executes the manufacturing activities, and the cyber part consists of the embedded system, which is a combination of input/output peripheral devices, computer processes, and computer memory [1]. PPC is a tool which helps in integrating and coordinating all the manufacturing activities in a manufacturing system. The production plan handles the materials planning, capacity planning, and operations scheduling, and the control portion oversees the actual production process to meet the production targets. The main aim of production planning and control is to minimize direct and indirect costs [2]. Maintenance management is the process of maintaining a firm's assets and resources. The main

DOI: 10.1201/9781003186670-8

purpose of maintenance management is to make sure that production runs in an efficient way and that the assets of a firm are used effectively [3].

In this context, the industries need CPS proficiencies for improving the usage of resources and increasing operator safety [4]. The integration of CPSs with PPC and MM helps industries fulfill different needs, such as efficient systems, reduction in system building and operational cost, and development of new innovative system capabilities, and it has mostly been recognized in the manufacturing, energy, and medical domains [5, 6]. Among various maintenance strategies, the condition-based strategy is dependent on the present condition, and consequently it needs to determine timing of necessities, which can be predicted with the help of predictive maintenance techniques at an early stage [7–8]. These maintenance techniques help in improving several challenges that affect Flexible System (FS) efficiency and performance in view of the breakdown of machines, maintenance issues, sudden interruptions due to natural characteristics, and so on [9].

The challenges of integrating a CPS for a manufacturing system have been observed from four viewpoints: improving the production, reconfiguration, information technology, and standardization. A 5C architecture with five levels, connection, conversion, cyber, cognition, and configuration, was developed by [10] to overcome some of the challenges mentioned previously. Here, accurate and reliable measurements of flexible systems can be obtained by connecting various sensors to the units, and this is done by the connection level. The measurement of data and useful information conversion are taken care of by the conversion level. More data can be obtained by connecting sensors to a greater number of machines, which is the purpose of the cyber level. Statistics and visual information to assist users to make decisions will be handled by the cognition level. Finally, feedback to the physical system according to the decisions made will be done by the configuration level.

According to the authors' knowledge, there are limited frameworks and approaches available in the context of integration of CPS across the product life cycle (PLC) [11]. In this chapter, we propose an integrated CPS with traditional PPC and MM for several flexible configurations that can cater to the needs of recent production industries. This chapter also concerns how Industry 4.0 integrates CPSs regarding maintenance activities and various needs for a company to attain the ideal factory. Finally, we provide the main advantages of integrating the CPS approach with PPC and MM that can increase machine availability, increase operator safety, improve product quality, and reduce maintenance cost.

8.2   LITERATURE REVIEW

The discussion in this section focuses on detailed literature about the CPS approach with PPC and MM as well as several challenges that affect system efficiency and the performance of realistic flexible configuration systems. CPSs became more popular in the context of the fourth industrial revolution (Industry 4.0). The main drivers for the development of CPS are security, competitiveness, social needs, and so on to reduce development costs [10] and time for improvement in designing products to make systems safer, increase productivity, and reduce maintenance cost. The relation between the designed product and manufacturing system plays a key role in the evolution of Industry 4.0 [12–16]. For building a CPS, an 8C architecture considering 3C facets along with 5C architecture that provide guidelines for a smart factory is proposed in [10]. The 5C architecture consists of 5 levels: connection, conversion, cyber, cognition, and configuration. The integration of the CPS approach with the production plan and MM of flexible configurations is important and can improve productivity [2].

PPC is planning for the production and manufacturing of various modules in an industry. Generally, a shorter PLC and the challenges faced by employees as a result of technological changes require an upgrade of their practice-related training and qualifications. Given this situation, the cost objectives are influenced by numerous interactive mechanisms. Decisions need to be made in the framework of PPC and targeted, as these objectives have to consider technical considerations. From the past literature, it is clear that CPSs in view of PPC are an advantage in the case of cost reduction [16]. Similarly, [17]'s investigation has shown that production planning is essential for manufacturing systems for reducing the overall cost. [18] briefly reviewed the technologies that contribute to take a production plan to a higher level by considering the Industry 4.0 tools which impact production processes particular to CPS. Along with that, [19] presented a method for a production and maintenance plan for a manufacturing system to minimize cost and maximize reliability.

The past literature has shown that inadequate maintenance practices also affect the industry's competitiveness by reducing the reliability of production facilities and lowering equipment availability [20]. To solve these mentioned problems, industrial system maintenance is an important part of asset management strategy that aims to maintain better levels of efficiency [3]. Generally, maintenance will lead to the monitoring of physical processes with the help of sensors, and it is a basic function of CPSs. It has been identified how Industry 4.0 integrates CPSs regarding maintenance management and the requirements for industries to attain the ideal smart factory. Thus, the impact of maintenance is mainly on profitability and productivity, which are the two most important business performance aspects. Additionally, many industries are seeking to facilitate performance assets and create a safer, more sustainable environment with the help of better asset management strategies. Moreover, the industry may face various challenges with the integration of CPSs into the manufacturing industry, such as data protection, data security, strategic planning, and so on [21]. [22] aimed to review the literature on CPSs for manufacturing the fourth industrial revolution for a complete understanding of its challenges and various techniques used in this domain.

8.3 PROPOSED FRAMEWORK FOR INTEGRATED CYBER-PHYSICAL SYSTEMS

This section presents the proposed framework for the integration of cyber-physical systems, with PPC and MM for flexible systems shown in Figure 8.1. The framework has been developed for

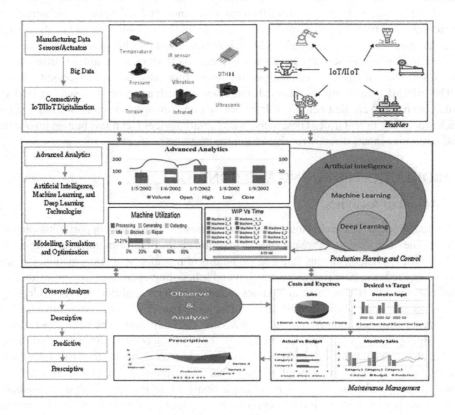

FIGURE 8.1 Framework for integrated cyber-physical systems for flexible systems.

understanding of CPS integration aimed at smart data collection and IoT/IIoT digitalization, along with a production plan for cost minimization and a maintenance plan for system safety and better efficiency.

8.4 ENABLERS

8.4.1 MANUFACTURING DATA COLLECTION FROM SENSORS/ACTUATORS

Integration of a CPS involves a development process which includes various devices such as sensors and actuators in the initial stage. Today, a large amount of data can be collected from these sensors and actuators. This mammoth amount of data retrieved from various industries needs to be filtered, segregated, and monitored for carrying out operations progressively; further, the data help in better decisions, future predictions, and discovery of next-generation innovative technologies. With the integration of the CPS, starting from data collection, data were characterized by large diversity and dynamics. Generally, these data are difficult to handle with the help of traditional data management tools. However, intellectual property analytics methods such as artificial intelligence (AI), machine learning (ML), and advanced data analytics depend on the quality and quantity to support building a better CPS. The mentioned technologies are acting as catalysts for the emergence of recent self-controlled manufacturing [23].

8.4.2 IoT/IIoT

The previously mentioned manufacturing data are received from the intelligent devices, namely the IoT and IIoT. The main difference between the IoT and IIoT is that the IoT works for the consumer's convenience, and the IIoT works for better efficiency and safety in manufacturing facilities. In general, the IIoT contributes to industries through its technology for developing industrial devices, which are industrial robots, sensors, and actuators connected for monitoring and analyzing the system and changing data easily and quickly [24]. It is believed that the IIoT plays a major role in developing CPSs in view of Industry 4.0. In this context, experts from major industries are exploring the possibilities of integrating these technologies for connecting network devices to the CPS environment. The IoT/IIoT consists of web-enabled smart devices such as processors, sensors, and various communication hardware that help to collect data in a uninterrupted, responsive, and fast manner for quick processing in real time. These sensors and smart devices mean less human involvement in FSs and customize inputs directly received from the customers [25, 26].

8.5 PRODUCTION PLANNING AND CONTROL

8.5.1 ADVANCED ANALYTICS

With the emergence of new technologies, many industries are facing new opportunities and several challenges. The traditional tools are not enough to capture enough information in this digital era due to the difficulties in heterogeneous data and mountains of manufacturing information. Particularly, when the data are huge, handling of sequencing and scheduling operations in production planning is a challenging task. Hence, most companies are shifting their traditional ways of dealing with data to the broader applicability and flexibility of advanced analytics. Integrating advanced analytics with latest communication technologies allows the CPS platform to be better operationalized, faster, flexible, efficient, and effective. In addition, these advanced analytics help in predicting the future (enabling organizations to make data-driven decisions), reducing the risks (avoiding making costly and risky decisions depending upon inaccurate predictions), anticipating and solving problems (based on likelihood, it can prescribe actions). More in-depth analysis is possible with advanced analytics methods with the integration of data mining, cohort analysis, cluster analysis, artificial intelligence, machine learning, and deep learning (DL).

8.5.2 ARTIFICIAL INTELLIGENCE, MACHINE LEARNING, AND DEEP LEARNING

Several works have proved that integration of AI technologies with CPSs plays a vital role in recent industrial applications. In the context of production planning and control, selection of the most suitable sensor for a particular operation on a particular machine provides greater benefits in terms of improving the productivity, reducing the cost and time, and so on. Further, self-learning–based techniques with respect to ML provide greater chances to improve the classification and clustering of mammoth data generated by geographically distributed enterprises. For example, for a case of flexible systems, actions can be taken from learning as outputs. A statistical learning algorithm may be used in a FS to automatically learn and try to improve without maintenance manager intervention. On the other side, DL learns from its experience, but large data need to be provided as the input with a large number of features. DL provides a high class of models which can approximate any function. These attributes are helpful when DL methods apply in cyber-physical systems because of the large amount of data collection from the CPS. DL algorithms help in using security-related applications in the context of CPSs and also can be used to improve the generalization capability of DL-based CPS security applications. The major areas where DL can be applied in CPSs are malware detection, threat hunting, intrusion detection, prevention of blackouts, vulnerability detection, and so on for security-related purposes in CPSs [24]

8.5.3 MODELING, SIMULATION, AND OPTIMIZATION

In production planning, different dynamic elements impact the movement of material in the manufacturing environment, and it is becoming complex for multi-product assembling plants due to various factors. Analyzing these factors can be a complex problem, and it requires modelling and simulation tools with respect to PPC in view of CPSs. It has been identified from the literature that generally small and medium enterprises (SMEs) have insufficient resources for collecting a huge amount of data and don't know how to process it effectively. Even SMEs fail in implementing various techniques and tools to analyze the data due to the higher cost of simulation or optimization tools and the lack of personal and specific knowledge. To reduce the burden in utilizing collected data for production planning, if the SMEs use a model, simulation and optimization approaches with respect to PPC will be helpful due to complex labor-intensive manufacturing in the view of integration of CPSs. Soft computing technique approaches for modeling, simulation, and recognition in data relationships, with the genetic algorithm for optimization of manufacturing resource configuration, can be used, and as been mentioned in [27]. Similarly, the processing and analytic layer is generally used to process data with the help of simulation for dimensional modeling algorithms. Using reports, graphical representation in a dashboard visualization could be generated for real-time monitoring purposes.

8.6 MAINTENANCE MANAGEMENT

8.6.1 OBSERVE/ANALYZE

Smart manufacturing describes fully integrated systems to meet tomorrow's real-time demand. Smart manufacturing is a technology-driven method that uses an internet connection and hardware approach to observe various production processes and support workers using novel methods of human–computer interaction. Today, newly developed industries are deploying more intelligent and smart technologies to observe/analyze manufacturing systems' health conditions, and the results show a 17–20% increase in productivity [28, 29]. Generally, smart manufacturing technologies include data analytics, collaborative robotics, the Internet of Things, additive manufacturing, and so on. These kinds of observations in a FS by collecting the data and analyzing them can be helpful in the context of analytic techniques such as descriptive, predictive, and prescriptive, as discussed in the following.

8.6.2 DESCRIPTIVE ANALYTICS

The arrangement of a present FS with the CPS could be subjected to a model of a production system, predicting plant information and solving various issues effectively. When solving various issues, the descriptive-analytic technique does precisely what the name implies: it describes the raw data as something that can be interpreted by humans. Nowadays, 90% of organizations use descriptive analytics, which is the easiest and most basic form of analytics. Generally, the easiest way to define descriptive analysis is as answers to questions about what happened in the past.

8.6.3 PREDICTIVE ANALYTICS

Among various data science techniques, the predictive analytic technique expands the quality and quantity of various complex data relevant to the manufacturing system health status. The predictive technique could provide information about various opportunities for understanding and control of temperature, pressure, moisture, and so on in a FS that affect the system's health conditions. Predictive analytics analyzes past information accurately and predicts what could happen in the future. This analysis helps in setting goals with effective planning and prevention expectations for a FS. Generally, the predictive analytic technique is to study data to find the answers to the question of what could happen in the future based on the past information.

8.6.4 PRESCRIPTIVE ANALYTICS

Nowadays, the prescriptive analytic technique is essential for smart and flexible production processes due to the increase in complexity, and traditional maintenance techniques are unable to fulfill production requirements. The development of predictive analytic techniques in manufacturing industries can reduce the production and process costs up to 30% and the breakdown of units up to 75% in comparison to traditional preventive techniques. However, with the digitalization of the manufacturing industry, a new era is emerging in the field of manufacturing systems maintenance, so-called prescriptive maintenance. On the other hand, prescriptive maintenance optimizes the production plan and schedule to meet requirements. A prescriptive analytics solution can recommend the optimal action plan, likely through a combination of mathematical algorithms, ML, and AI for better outcomes from FS.

8.7 FLEXIBLE CONFIGURATIONS

The flexible real-time configurations shown in Figure 8.2 demonstrate the integration of the CPS approach in view of PPC and MM that has been proposed in this chapter. The FS is classified into four types: one degree of flexibility, two degrees of flexibility, semi-flexible, and fully flexible configurations. The FS consists of N of identical units, which perform similar operations and simultaneously process the given number of jobs. Here, when a job arrives, it can be given to any functional unit to complete in this configuration. (1,1), (1,2), . . ., (m,n) indicate identical units in which the connection from unit to unit with the arrows shows one degree of flexibility. If any unit fails, the remaining jobs can be assigned to the adjacent machine, that is, unit (1,2), depending upon the unit availability. The connection between the units indicated by the black and blue colors shows the two degrees of flexibility environment, in which the availability of the units is greater compared to the one degree of flexibility configuration. If machine (1,1) fails, then the pending jobs can be assigned to units (1,2) and (2,2). The connection between the units indicated by black, blue, and orange shows the semi-flexible configuration, and black, blue, orange, and green show the fully flexible configuration, where the number of units available is greater compared to the one and two degrees of flexibility configurations.

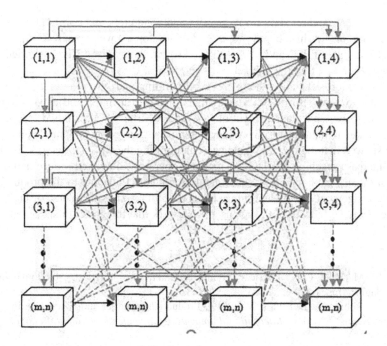

FIGURE 8.2 Flexible configurations.

Real-time small-scale industries consist of a one degree of flexibility configuration which is about 50 identical machines performing a similar operation. The major applications of a one-degree of flexibility configuration are canned goods, over-the-counter drugs, household appliances, and so on. Medium-scale industries consist of two degrees of flexibility and semi-flexible configurations, which are 50–200 identical machines performing a similar operation. The major applications of two degrees of flexibility and semi-flexible configurations are computer chips, clothing, and so on. Large-scale industries consist of fully flexible configurations, which are close to 200 identical machines. The major applications of fully flexible configurations are for complex products such as the manufacturing of gears, and so on.

8.8 MANAGERIAL IMPLICATIONS

The integration of CPSs includes computers communicating with the physical processes in a FS and mediating the way they interact with the physical world. The integrated CPS with the FS helps to observe the production plan and maintenance characteristics of a unit system, and it allows the operator remote control of the FS to analyze, configure, monitor, and diagnose the CPS's abilities. The CPS can analyze the flexible environment in which the unit operates and request interventions from the predictive maintenance-based condition analysis. The CPS helps in human–machine interface technology and creates a user-friendly environment for the maintainer for easy and quick access to unit information by implementing their ideas and activities within a short time. This technology also provides information about the history of the fault system before any maintenance is required by the maintainer.

8.9 CONCLUSIONS

CPS is expected to play a key role in the design and development of future manufacturing systems with advanced capabilities that far exceed the current level of functionality, reliability, and cybersecurity. A CPS captures data using sensors and computers and communicates alerts to the manufacturing systems. In this chapter, we aimed to develop a novel framework for the integration of a CPS to a FS in light of PPC and MM. As stated so far, there exist different challenges that have been discussed when tackling the subject of integration of CPSs on FSs for Industry 4.0. It leads to such benefits as more efficient systems, reduced building and system operation costs, and safer and more reliable systems. So far, CPS as a smart manufacturing-related approach and analysis, visualization, and new tendencies in light of PPC and MM have been studied. Along with that, a few examples related to manufacturing systems integration and its working has been discussed. The investigation showed that the main advantages of integrating the CPS approach with PPC and MM is that it can increase machine availability, increase operator safety, and reduce maintenance cost. The investigation also explained how such integration helps contribute to the production plan and maintenance management for an ideal future factory. Summarizing the related discussion leads to a focus on aspects facing Industry 4.0, such as a framework that integrates on a FS.

REFERENCES

1. He, K., & Jin, M. (2016). Cyber-Physical Systems for Maintenance in Industry 4.0. Jonkoping University, School of Engineering, 1–64.
2. Rossit, D. A., Tohme, F., & Frutos, M. (2019). Production planning and scheduling in cyber-physical production systems: A review. *International Journal of Computer Integrated Manufacturing*, 32(4–5), 385–395.
3. Schreiber, M., Vernickel, K., Richter, C., & Reinhart, G. (2019). Integrated production and maintenance planning in cyber-physical production systems. *Procedia CIRP*, 79, 534–539.
4. Neal, A. D., Sharpe, R. G., van Lopik, K., Tribe, J., Goodall, P., Lugo, H., . . . & West, A. A. (2021). The potential of Industry 4.0 cyber physical system to improve quality assurance: An automotive case study for wash monitoring of returnable transit items. *CIRP Journal of Manufacturing Science and Technology*, 32, 461–475.
5. Okolie, S. O., Kuyoro, S. O., & Ohwo, O. B. (2018). Emerging cyber-physical systems: An overview. *International Journal of Scientific Research in Computer Science Engineering and Information Technology*, 306–316.
6. Yaacoub, J. P. A., Salman, O., Noura, H. N., Kaaniche, N., Chehab, A., & Malli, M. (2020). Cyber-physical systems security: Limitations, issues and future trends. *Microprocessors and Microsystems*, 77, 103201.
7. Yildirim, M., Gebraeel, N. Z., & Sun, X. A. (2019). Leveraging predictive analytics to control and coordinate operations, asset loading, and maintenance. *IEEE Transactions on Power Systems*, 34(6), 4279–4290.
8. Von Birgelen, A., Buratti, D., Mager, J., & Niggemann, O. (2018). Self-organizing maps for anomaly localization and predictive maintenance in cyber-physical production systems. *Procedia CIRP*, 72, 480–485.
9. Shagluf, A., Longstaff, A. P., & Fletcher, S. (2014). Maintenance strategies to reduce downtime due to machine positional errors. In *Proceedings of Maintenance Performance Measurement and Management (MPMM) Conference 2014*. Imprensa da Universidade de Coimbra.
10. Jiang, J. R. (2018). An improved cyber-physical systems architecture for Industry 4.0 smart factories. *Advances in Mechanical Engineering*, 10(6), 1687814018784192.
11. Barthelmey, A., Störkle, D., Kuhlenkötter, B., & Deuse, J. (2014). Cyber physical systems for life cycle continuous technical documentation of manufacturing facilities. *Procedia Cirp*, 17, 207–211.
12. Cui, X. (2021). Cyber-Physical System (CPS) architecture for real-time water sustainability management in manufacturing industry. *Procedia CIRP*, 99, 543–548.
13. Hehenberger, P., Vogel-Heuser, B., Bradley, D., Eynard, B., Tomiyama, T., & Achiche, S. (2016). Design, modelling, simulation and integration of cyber physical systems: Methods and applications. *Computers in Industry*, 82, 273–289.
14. Yan, J., Zhang, M., & Fu, Z. (2019). An intralogistics-oriented Cyber-Physical System for workshop in the context of Industry 4.0. *Procedia Manufacturing*, 35, 1178–1183.
15. Saldivar, A. A. F., Li, Y., Chen, W. N., Zhan, Z. H., Zhang, J., & Chen, L. Y. (2015, September). Industry 4.0 with cyber-physical integration: A design and manufacture perspective. In *2015 21st International Conference on Automation and Computing (ICAC)* (pp. 1–6). IEEE.

16. Seitz, K. F., & Nyhuis, P. (2015). Cyber-physical production systems combined with logistic models—a learning factory concept for an improved production planning and control. *Procedia Cirp, 32*, 92–97.
17. Meissner, H., & Aurich, J. C. (2019). Implications of cyber-physical production systems on integrated process planning and scheduling. *Procedia Manufacturing, 28*, 167–173.
18. Ebrahimipour, V., Najjarbashi, A., & Sheikhalishahi, M. (2015). Multi-objective modeling for preventive maintenance scheduling in a multiple production line. *Journal of Intelligent Manufacturing, 26*(1), 111–122.
19. Boyes, H., Hallaq, B., Cunningham, J., & Watson, T. (2018). The industrial internet of things (IIoT): An analysis framework. *Computers in Industry, 101*, 1–12.
20. Kanisuru, A. M. (2017). Sustainable maintenance practices and skills for competitive production system. In *Skills Development for Sustainable Manufacturing*. IntechOpen.
21. Giaimo, F., Andrade, H., & Berger, C. (2020). Continuous experimentation and the cyber—physical systems challenge: An overview of the literature and the industrial perspective. *Journal of Systems and Software, 170*, 110781.
22. Dafflon, B., Moalla, N., & Ouzrout, Y. (2021). The challenges, approaches, and used techniques of CPS for manufacturing in Industry 4.0: A literature review. *The International Journal of Advanced Manufacturing Technology*, 1–18.
23. Radanliev, P., De Roure, D., Van Kleek, M., Santos, O., & Ani, U. (2020). Artificial intelligence in cyber physical systems. *AI & Society*, 1–14.
24. Wickramasinghe, C. S., Marino, D. L., Amarasinghe, K., & Manic, M. (2018, October). Generalization of deep learning for cyber-physical system security: A survey. In *IECON 2018–44th Annual Conference of the IEEE Industrial Electronics Society* (pp. 745–751). IEEE.
25. Pivoto, D. G., de Almeida, L. F., da Rosa Righi, R., Rodrigues, J. J. P. C., Lugli, A. B., & Alberti, A. M. (2021). Cyber-physical systems architectures for industrial internet of things applications in Industry 4.0: A literature review. *Journal of Manufacturing Systems, 58*, 176–192.
26. Phuyal, S., Bista, D., & Bista, R. (2020). Challenges, opportunities and future directions of smart manufacturing: A state of art review. *Sustainable Futures, 2*, 100023.
27. Khan, W. Z., Rehman, M. H., Zangoti, H. M., Afzal, M. K., Armi, N., & Salah, K. (2020). Industrial internet of things: Recent advances, enabling technologies and open challenges. *Computers & Electrical Engineering, 81*, 106522.
28. Teerasoponpong, S., & Sopadang, A. (2021). A simulation-optimization approach for adaptive manufacturing capacity planning in small and medium-sized enterprises. *Expert Systems with Applications, 168*, 114451.
29. Tao, F., Qi, Q., Wang, L., & Nee, A. Y. C. (2019). Digital twins and cyber-physical systems toward smart manufacturing and Industry 4.0: Correlation and comparison. *Engineering, 5*(4), 653–661.

9 Wearables to Improve Efficiency, Productivity, and Safety of Operations

K. Balamurugan, T.P. Latchoumi, and T.P. Ezhilarasi

CONTENTS

9.1 Introduction 75
9.2 Related Works 77
9.3 Sensors and Devices for Mobile Technology 78
9.4 In Construction Engineering 78
9.5 In Other Industries... 78
9.6 Fall Protection for Highway and Bridge Workers 79
9.7 Objectives 80
9.8 Operable: The Proposed Wearable 80
9.9 Movement Characterization 81
9.10 Physiological Response as a Measure of Safety and Efficiency 81
9.11 Methods.. 82
9.12 Results and Observations 82
9.13 Construction Safety and Health Hazards 82
9.14 Loading and Unloading Tasks 83
9.15 Low-Frequency Movement Characterization Algorithm 84
9.16 Conclusion... 87
References 87

9.1 INTRODUCTION

Professionals and scientists are worried about the high number of losses in the building industry. Construction was responsible for 899 of the 4,386 fatal incidents in the private sector in 2014, meaning that more than one in every five employees died as a result of construction [1, 2]. Construction workers are the most vulnerable to industrial accidents and illnesses [3]. The Occupational Safety and Health Authority (OSHA) has established and mandated safety procedures and services. Over the last decade, the number of deaths, injuries, and non-lethal illnesses in the construction industry has remained stable.

Since the building industry has such a high rate of fatal and non-fatal injuries, businesses are still looking for new ways to improve safety [4]. Since construction is a transient and competitive industry, organizations respond to rapid changes by storing, capturing, and disseminating new injury prevention techniques efficiently [5]. As a result, emerging innovations could be a candidate for safety advancement. Although technology has undeniably improved construction procedures, its application to customized construction safety monitoring has yet to be thoroughly investigated [6].

The building and construction industry, despite being one of the most influential, faces a variety of challenges. Because of the complex, transitory, and dangerous nature of construction activities, for example, safety management is especially difficult. According to figures, the building industry

DOI: 10.1201/9781003186670-9

causes more than 60,000 fatal accidents per year around the world. Staff and their families suffer a great deal as a result of these accidents, resulting in increased annual losses.

High efficiency and low productivity rates, like low safety results, are a universal challenge in the construction industry [7–9]. For example, because of efficiency gains, productivity rates in certain industries (such as manufacturing) have doubled. Over the last two decades, productivity in the building industry has remained relatively stagnant [10]. The construction industry's productivity deficit with the rest of the economy is projected to be worth $1.6 trillion [11].

Most studies aimed at increasing efficiency do not consider safety implications, and vice versa [12]. Unfortunately, in some cases, this limited viewpoint could have unintended consequences [13]. For example, [13] article claims that to meet production and efficiency goals, employees often prioritize safety methodologies [14, 15]. Furthermore, evidence suggests that when productivity pressures are high, risk-taking behavior increases [16]. As a result, preventing such accidents necessitates measures that benefit both protection and quality and productivity. To achieve this goal, the research presented centers on national safety problem currently confronting transportation agencies, taking quality, effectiveness, and other safety factors into account.

When road and bridge staff performs bridge maintenance on the bridge, fall safety is a problem [17]. These staff members rely heavily on existing railings for fall safety at lower levels when performing these tasks [18]. Several transportation agencies, including the North Carolina Department of Transportation (NCDOT), are working to reduce the danger of falls. The architecture and operational features of these fall defense systems vary. NCDOT conducted an initial examination into more than 50 of these Flow Shop Sequence Dependent Setup (FSSDs) as part of its groundbreaking efforts and agreed to pursue four options for widespread adoption [19, 20].

The current survey aims to evaluate the four different FSSDs in greater depth. The aim is to evaluate FSSDs holistically in terms of their protection, performance, and productivity benefits, rather than concentrating solely on safety. All transportation organizations looking to implement such fall safety policies will benefit from the survey results. Current research aims to promote efforts to improve the quality and productivity of road and bridge staff, in addition to resolving a national safety problem.

By the output rates and encouraging cooperation, the fourth industrial revolution was born to change the new industrial model and bring digitization into conventional factories. Industry 4.0 [21] is the name given to this new revolution. Industry 4.0 focuses on the IoT [22] and how it's being used in industrial networks to link objects, equipment, and humans in intelligent factories. Rising output rates and lowering costs help collaborative tasks. This is accomplished by the use of cyber-physical systems (CPSs) [23], which are described as using sensors together information and run cognitive algorithms over the Internet.

This latest paradigm places a strong emphasis on sustainability. To address Industry 4.0's biggest challenge not just ecological but also public sustainability is encouraged. New CPSs should be converted into employee-centered human architectures that ensure protection, well-being, and comfort to achieve the best possible operational environment and output. The growth of the wearable market plays a role of a key marker of worldwide achievement, allowing for the development of smarter devices by integrating sensors and actuators. Advancements like Bluetooth low energy (BLE) will take the direction of IoT standardization as a result of the need for a further flexible alternative, and it is leading an ongoing industrial revolution in which diffusion capability will play a key role in the current approval of BLE web topology. Meanwhile, accelerometers have started an IoT wave that enables gestural devices to interact with humans and machines more effectively.

Massive amounts of data are generated by these devices and must be processed and filtered to obtain exact data for the worldwide challenge called big data. This chapter addresses the major challenges of Industry 4.0 used in OperaBLE. By immersing workers in CPSs while taking into account personal care facilities, this prototype work strip based on BLE aims to incorporate digitalization in smart factories and improve working conditions. It is a modern wearable that enhances safety in engineering settings by anticipating unsafe situations and workplace injuries. Because of the wide range of creative technologies for the Industrial Internet of Things (IIoT), low-power devices would have a positive impact on the work. OperaBLE utilizes its learning capacity to make note of operations and improve the power value chain. It specializes in movement analysis methods and heart rate analyses. The basic goals in this chapter are to define an index and evaluate the characteristics of portable technology, as well as the resulting data, which can be used to forecast protection and construction management practices.

9.2 RELATED WORKS

Wearables with accelerometers are particularly useful for keeping track of private actions and power utilization. Last year, the number of industrial wearables increased (mostly wrist bands labeled as intelligent bands). In terms of research practice, there are many categories: medicine and recovery, movement awareness, older people, athletics, location, and jobs. In terms of medical aspects, [24] proposes a systematic scheme to prevent nominal infection, which includes sensors and bracelets with accelerometers. Other intelligent bands for muscle movement identification and physiologic signals have also been suggested [25] to help patients with obstructive sleep apnea and to diagnose Parkinson's disease. Furthermore, accelerometer-based systems are often used in rehabilitation.

When it comes to accelerometer-based applications, there is a lot of research into gesture and motor recognition. A laptop or even a smart phone can be used for this. In [26], an accelerometer for movement monitoring was presented for detecting deteriorating balance, and finally, it was used to detect certain predefined actions in youthful and aged adults. The authors also created and introduced an integrated method to estimate pedestrian walking positions in [27].

Finally, several studies on the use of accelerometers in the workplace have concentrated on improving workplace safety in the construction industry. Industrial accelerometers were used to calculate trunk and longitude in a flexion angle in a lab setting in [28], and this research is continual in a real-world setting [29]. In cooperation, studies looked at data from commercial accelerometers introduces a construction worker method concept based on the supervised decomposition by the sensor movement [30, 31]. Other research is based on different types of employees; for example, the authors of [32] investigated the use of accelerometers in assessing a variety of professionals. It has not yet been considered for industrial settings, where it could play a significant role in the emerging Industry 4.0.

Workers are often exposed to health hazards and take risks during the construction procedure owing to hazardous work situations on sites. Protection during construction has traditionally been treated and assessed in a thoughtless way as a reaction to negative damage trends. Dynamic monitoring of physiological labor information through portable tools, on the other hand, allows measurement of pulse rate, respiration, and attitude [33]. This section includes a summary of related literature on security performance monitoring, wearable device and sensor types, and wearable technology applications in construction and other industries.

The majority of traditional methods for calculating security competence capacity are labor intensive and based on biased assessment [34]. These approaches depend on the manual processing of vast volumes of electronic data; as an effect, information is gathered rarely (e.g. monthly) and in the event of an accident [35]. These methods are costly, prone to loss of data, and result in datasets that are too limited to efficiently manage works. Although it overcomes the weaknesses of manual processing, automated security monitoring is perhaps the most efficient method for reliable as well as consistent tracking of safety performance across construction projects [36].

Many real-time project management approaches have clear safety applications. Safety health oversight aims to ensure that the safety activities of construction workers are effectively treated and controlled in compliance with existing safety plans and standards. Standard industrial control systems are unfortunately impractical in construction due to the transient existence of construction sites and project organization [37]. Automatic monitoring of human positions and trajectories can be useful for a variety of engineering applications, including defense, security, and process analysis [38].

Wearable devices provide continuous tracking of a wide range of vital signs, as well as preventive measures for employees with high-risk physical conditions.

9.3 SENSORS AND DEVICES FOR MOBILE TECHNOLOGY

Magnetic fields, an ultra-wide band (UWB), radio-frequency identification (RFID), microwave, infrared, wireless, thermal imaging, global positioning systems (GPSs), video, projectors, static cameras, angiograms, and electromyography are some of the devices used in wearable technologies. A network of body sensors is created by sensors, such as galvanic skin response (GSR) sensors, actuators, propellers, and detectors.

9.4 IN CONSTRUCTION ENGINEERING

In comparison to other sectors, the use of wearable technology and materials in construction is still in its early stages. In reality, the building industry has only a few known cases of wearable device application [39]. The assessment of a method for evaluating sensitivity identification and warning systems to facilitate protection on building sites is one of a few apps. By constantly gathering information on the worksite and sensing environmental risks, as well as the distance of employees to hazardous locations, hands-free devices have been used to track employees to maximize their spatial awareness. Part of the reason for their shortage of mass acceptance is a scarcity of credible evidence to back up the potential benefits.

Smart phones have commonly been used by the building sector to view and exchange construction information from external construction sites. Even though the building industry may have been slow in adopting developments in accessibility and automation devices, as well as some other innovations that can boost efficiency, smart devices may reveal opportunities for growth [40]. As a result, there seem to be a lot of opportunities for the construction sector to use wearable devices for customized safety.

9.5 IN OTHER INDUSTRIES

Medicare, construction, mining, and telecommunications have all made use of wearable technology. These kinds of advance results showed success, and both companies and business experts are working to improve them and the benefits of their initial developments. Because of the advent to flower-power computer systems and cheaper detectors, wearable techniques are mostly being used in public health to facilitate physical activities [41]. Individual health monitoring systems using micro semi-sensors to capture and analyze human physiological parameters have made significant advances in computing technologies. A variety of small portable detectors are available [42]. Innovation in mini devices and wireless techniques, a future innovation of surveillance methods, is now possible.

Remote monitoring of patients also helps patients monitor their well-being while avoiding unnecessary doctor visits. Several companies have developed body sensor networks for commercialization based on the pioneering work done by researchers at the National Aeronautics and Space Administration's Jet Propulsion Laboratory (NASA-JPL) [43]. Such devices offer fitness trackers, with the action carried out through a wireless system, and health monitoring systems that monitor information such as blood pressure, movement, oxygenation, temperature, and stance in real time to reduce medical costs and enhance efficiency [44]. Among the applications designed to enhance information transfer, performance, and safety in industry operation are inhibited contact, client care, remote management, and a stock management [45].

Portable technologies are commonly used in sports to track actions via seamless and discrete readings [46]. Wearable machines such as GPS wristwatches, blood pressure detectors, and fitness trackers are widely used to collect real-time data about outcomes. Professionals use wearable technology in a variety of devices to track not only their performance but also their protection [47]. Players in the National Football League (NFL) have concussion devices mounted in their helmets, for example. Smart compression shirts have been used in Major League Baseball (MLB) to test a pitcher' effectiveness by measuring arm movement and technique. Golfers use wearable bracelet GPS sports watches to improve their swing mechanics during workouts [48, 49].

Police officers, ships' crews, and paramedics test portable technologies to provide communication support remotely in security applications and input with the ability to access knowledge when conducting vital tasks without having house your hands [50]. Lighting systems and safety equipment are also used to improve visibility and draw attention to personal protection.

9.6 FALL PROTECTION FOR HIGHWAY AND BRIDGE WORKERS

While such studies have been done on falling, mostly in the construction sector as a whole, there has been little done specifically on falls for roads and bridge employees. Nevertheless, road and bridge employees, like those in the specific building industry, sustain and is proportionately large number of fall-related accidents. Falls are especially dangerous for road and bridge staff who work at heights, such as those on bridge decks. Each year, for example, approximately 3,000 fall-related accidents in road and bridge staff are registered. Furthermore, research shows that more than 80% of fatalities happen while on-the-deck bridge projects are in progress.

When operating at peaks or on endplates, there are many options for preventing falls. The least successful solution, according to the structure of fall security measures, as seen in Figure 9.1, is the use of admin safety controls that depend on staff to follow accepted safety methods and processes (e.g., maintain a safe distance from incompliant guardrail). Sadly, between road and bridge staff, it is the most commonly ignored fall safety method. One of the most powerful strategies, on the other hand, is to eliminate the risk of falling. Another way to implement this is to construct bridge safety barriers that provide adequate security, obviating any kind of additional security measures.

Tragically, many bridge fences weren't built to provide employees with adequate falling safety. Furthermore, under prevailing budget limitations, altering or removing existing bridge safety barriers is not technically possible. Some fall-prevention strategies have included fall-arrest as well as fall-restraint devices (e.g. lanyards). Regrettably, even though such fall safety technologies are largely used in the construction management sector, road and bridge employees are frequently prevented from using them due to financial restrictions. These limitations are typically in effect to prohibit the use of hooks that damage bridge materials to protect the public infrastructure's structural strength.

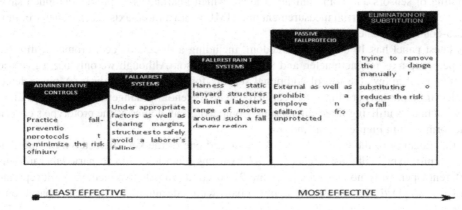

FIGURE 9.1 Hierarchy of fall protection controls.

The use of passive FSSDs, like the models studied in this study, seems to be another efficient and appropriate method of preventing falls. However, as previously mentioned area variety of FSSDs that could be on the marketplace. This involves comparing the various options and selecting the options that provide the maximum benefits in terms of protection, reliability, and usability.

9.7 OBJECTIVES

The information essential to produce this comparative analysis was collected appropriately utilizing two main approaches. The first approach included information gathering utilizing wearable devices in which laborers are involved in 96 field tests. Alternative FSSDs were used in field conditions for routine tasks such as falling safety device unloading, installation, deployment, and dismantlement. The Zephyr Bioharness 3 (Zephyr Technology, Annapolis, Maryland), which is a wearable system that has already been verified in a large number of studies, was used to collect data and information.

The second method involved conducting a utility study of the alternative FSSDs using a survey method. This entailed gathering feedback from workers on how they handled and operated FSSDs during field tests. But apart from collecting information on FSSD protection, performance, and profitability advantages, the usefulness review also obtained information on the instruments' reliability as encountered by employees throughout field trials.

The emphasis of the field testing was on several metrics reflective of integrity, performance, and efficiency since each of the four FSSDs fully resolved the falling security problem by providing a diffusion length that exceeded 99 cm (39 in.). The various metrics of protection, reliability, and profitability being used are addressed in detail in the following subsections.

9.8 OPERABLE: THE PROPOSED WEARABLE

This study examines a wearable device designed to meet IoT requirements in manufacturing technologies. It's a self-contained computer that can learn a person's behavioral habits and improve their operating consequences as a result. While unique hardware is suggested in this research to introduce a testable watch, OperaBLE's significant contributions are in its accessibility and, as a result, in the proposed algorithms. Different distinctions have been employed for the production of OperaBLE, and the first is the efficiency for Industry 4.0 technologies, as well as the second, its long-term activity. As a result, in the even conflict, proper performance takes precedence over energy consumption. OperaBLE's main challenge is to ensure that its features can be implemented in just about any lightweight IoT system with such a high level of reliability. Due to the decreased micro-controller functioning frequency that characterizes some small IoT devices and the successful handling of sensors and communication units, which guarantees to provide as much save in energy. We designed an inertial measurement unit (IMU) with a three-axis accelerometer built into the micro-controller framework.

This latest panel has 10 degrees of freedom, including a three-axis accelerometer, three-axis gyroscope, three-axis magnetometer, and temperature sensor. Although we only use a gyroscope in this section, it allows the use of additional sensors for any further analysis. A pulse detector is included in monitor operating heartbeat and detects potentially dangerous conditions. Finally, we modeled houses utilizing three-dimensional computer technologies as the prototype's covering, with the form seen in Figures 9.2(a) and (b).

The functionality of the OperaBLE model is indeed one of the key characteristics, and that is why the features to be built must be checked under various conditions. Our scenario laboratory used two different OperaBLE models (seen in Figure 9.2) used to evaluate two methods built separately. OperaBLE is a hybrid of low-cost prototyping board with a design algorithm that is based on the ability to adapt, low wattage activity, and sample rate.

(a) Exterior design: Operable with pulse sensor (b) Interior design: Operable
withaccelerometer

FIGURE 9.2 OperaBLE prototype used for experimentation.

9.9 MOVEMENT CHARACTERIZATION

This chapter goes through the algorithmic platform that allows for OperaBLE-defined motions. First and foremost, it is important to emphasize that motion detection using accelerometers is not a straightforward thing, because they provide direct acceleration data that must be collected, then interpreted properly. For example, distinguishing correct body motions and actions through gravitational force is a challenging task in which information via accelerometers is required to be interpreted by a challenging algorithm that necessitates extensive processing and predictive analysis. As stated earlier, one of the experiment's major accomplishments is the creation of a method formation detection using only a sampling rate of about 10 Hz, which is up to 10 times less than that used in traditional experiments. Such a feature offers significant benefits in terms of more effective battery use, with the following contributions:

(1) The motion characterization methodology can be used in a variety of devices, including those that are regulated by limited frequency microcontrollers, which use less power than large processing mechanisms. As a result, the algorithm's ability to analyze direct acceleration data and classify individual body language while using the fewest resources possible allows it to be implemented in compact and cost-effective devices, leading to more efficient usage of resources.

(2) Other than being effective in improving characterization with a smaller sample per action, the techniques themselves necessitate a low measurement operating frequency. Drop samples are not an integral energy-saving feature of OperaBLE, and they do help to reduce the amount of data collected, which reduces the amount of power required by accelerometers. The use of BLE digital technology allows for the optimization of the communications service module energy demand. The activity characterization methods process is based on the assumption that even if the content acquired were irrelevant to OperaBLE (which also pre-processes actual displacement information), no data are sent to the processing point. BLE allows for periods of sleep between data transmissions with minimal power usage, and so as a result of the proposed pre-processing stage, the quantities of information encapsulated and transmitted via OperaBLE are significantly decreased.

9.10 PHYSIOLOGICAL RESPONSE AS A MEASURE OF SAFETY AND EFFICIENCY

Overexertion-related accidents is a significant issue between road and bridge employees due to the physically strenuous nature of their employment. Tiredness and sheer fatigue are also common among these employees and are leading causes of the disproportion at risk of accidents. The portable technique uses effective and accurate data collection methods, whereas physical exertion is hard to monitor and quantify visibly. Since increased levels of physical burden will result in increased levels of potential risk (i.e., risk of overtraining, including inflammation accidents) as time progresses, rise in blood pressure is being used as a measurement of safety.

Furthermore, employee condition could be generally considered acceptable while skills required are low and excessive energy consumption is minimized, so an improvement in heartbeat is being used as a measure of progress. Employees would be able to work productively as well as for longer periods without being fatigued or exhausted, resulting in increased production costs.

Many of the tasks that road and bridge staff perform necessitate the use of uncomfortable body positions, including awkward poses. That's also especially true when employees are leaning, spinning, raising, and bending. Such practices may put an undue amount of strain on all these employees' neuromuscular systems, leading to spine fractures as well as other overexertion-related accidents. As a result, reducing any need to bend, turn, and carry during the working period will minimize the chance of such accidents among these employees.

9.11   METHODS

This chapter began with a summary of the current state of knowledge of mobile technology throughout businesses, followed by the codification of research, including specifications relevant to each applied technology device and detector, as well as a description of the technology's human performance consequences in alignment with dominant theory. Infrastructure safety and health challenges were examined in identifying the indicators that smart watches should acquire and analyze to assess and track protection results. An analysis of the literature showed four categories of observable safety efficiency metrics: physiological tracking, atmospheric detection, proximity tracking, and location tracking. A thorough examination of these four categories, including the protection performance indicators that can be assessed and tracked, as well as the relevant smart watch devices, including detectors, is provided. Furthermore, commercially available wearable devices were evaluated against performance expectations for practical customized wearable technology. Internet analysis was done to locate major wearable device producers and also recorded studies on wearable device applications around the world. We gathered and analyzed data about wearable gadgets via producer specifications manuals and released scientific papers and introduced and addressed a concept for integrating various wearable devices and detectors into the layout of practical wearable devices for customized protection measuring performance during development.

9.12   RESULTS AND OBSERVATIONS

According to the research methodology system, the findings of this analysis are summarized and addressed in various parts. Each study's key findings are discussed, along with the wearable computing devices and technologies that are believed will become the most effective for safety management tracking and development.

9.13   CONSTRUCTION SAFETY AND HEALTH HAZARDS

On any specific day, nearly 6 million workers are working at 272,000 construction projects throughout the United States. Each day, these employees are exposed to a variety of potential hazards, putting them at risk of being sick, injured, and even disabled for life. The construction sector has a greater mortality rate than that of the national median for all sectors in this group. Over the years, the number of injuries and illnesses on building sites has risen. Falling from a height caused by unsafe support beams or staircases, repetitive stress injury, heatstroke or heat exhaustion as a result of body temp increasing to hazardous levels, and being hit by moving machinery operating in close vicinity to employees are only a few of the possible safety and health risks for construction professionals. To have safety, the observable indicators should be connected/carried along with the workers. For effective implementation of wearable technology, compilation and evaluation of the

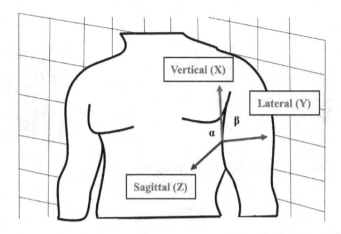

FIGURE 9.3 Motion captured by the wearable device.

measurements are needed to reduce the level of risk. The human safety facets can be implemented through special motion detecting devices fixed on the body of the worker, as shown in Figure 9.3.

With the support of NCDOT management, collaboration among road and bridge staff was sought in preparing for field tests. Provided that the use of FSSDs for highway construction is a relatively recent concept introduced by NCDOT, only 2 of 14 units in North Carolina have used them before. As a result, six personnel from of the two units were selected to take part in 96 field tests as a result of the frequency test.

When implementing one of its four FSSDs (i.e., six employees, four activities, four FSSDs, 96 field experiments), each of the workers performed four of the most-used techniques (i.e., Loading, Unloading, Deployment, and Disassembling). Moreover, while the participants were familiar with FSSDs in particular, they might have no previous knowledge of the particular FSSDs investigated in this review. The researchers only had previous knowledge of a custom-designed FSSD produced by NCDOT contractors, which was not included in the present study. This removed any questions about validity that may have harmed the study's findings due to prior knowledge.

The staff who participated in this research were all in good health and ranged in age from 30 to 60 years. The employees had between 6 and 20 years of professional experience in road and bridge infrastructure evaluation, renovation, and reconstruction. After the employees were informed of the costs involved with both field experiments, they signed a document consent form indicating readiness to participate in the research. The parts that follow explain the field and laboratory experimental design in light of the four methods that were used as related to the research initiative.

9.14 LOADING AND UNLOADING TASKS

When the FSSDs were implemented in operation, the loading and unloading of FSSDs became two processes that must be done regularly. Staff would have to take the FSSDs over to a vehicle, transfer them to the bridge site, then offload them to prepare for both scheduled work activities, for example. Employees also had to reload the FSSDs onto the vehicle and offload them in a container yard at the beginning of each working hour until the work shift activities were completed. Field tests included simulations of FSSD loading or unloading. To reduce the possibility of a traffic-related event, the simulation was carried out in a container yard where necessary.

The contributing employee was given a proper summary of portable sensors before starting one of the simulated field activities but was made to wear the device according to the maker's guidelines. After that, the employee was told to sit and relax on a bench beside the vehicle to get his or her pulse rate back to normal. Throughout this period, the researchers put the FSSDs at a range of 6.07 m (25 ft)

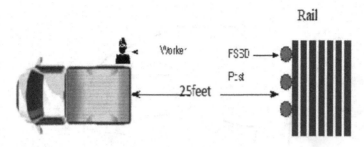

FIGURE 9.4   Placement of the FSSDs before the loading task.

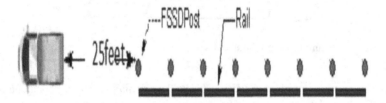

FIGURE 9.5   Placement of the FSSDs after the unloading task.

from either vehicle, as seen in Figure 9.4. Following the schedule, the involved employee was tasked with loading the FSSDs onto a vehicle, as is customary in bridge planning. It was necessary to do eight reports as part of the mission.

After loading was over, the employee was given a place to relax before his or her pulse rate stabilized. The employee was again given the responsibility of offloading the previously filled FSSDs from the vehicle and arranging the components, as shown in Figure 9.5. The proposal replicated the conventional method of positioning FSSDs across a bridge's fence in readiness for construction. Each of the staff was treated the same way for every one of the FSSDs throughout the present study.

After installation was performed, the employee was given another seated break to allow the pulse rate to stabilize. Following the break, the employee removed the fall safety pillars, which included softening, raising, and repositioning the posts, mostly on the bridge. The process was replicated for each of the staff members who participated.

9.15   LOW-FREQUENCY MOVEMENT CHARACTERIZATION ALGORITHM

The protocols carried out by our algorithm, the low-frequency movement characterization algorithm (LoMoCA), are presented in this chapter (the process diagram given in Figure 9.6 includes representative stages). OperaBLE's methodology is separated into two sections: pre-processing, then non-wearable computer deployment (processing). In terms of the smart interface, we created a battery-saving code that keeps the system in a sleep state when there is no noticeable activity. As OperaBLE detects the start of a new activity, the subroutine gathers information from the IMU's combined accelerometer and gyroscope and stores it in the system until the activity is completed.

This moving trigger activity is caused by a threshold that compares the previous and present samples in real time to see if there is a difference of more than 1 m/s^2 (acceleration) and 1 rad/s (angular acceleration). The computer goes into sleep mode when waiting for such a major acceleration and angular velocity in an instant to decide the conclusion of each action. The collection of information is fully achieved when no activity is detected throughout the period. The movement pattern is shown in Figure 9.7.

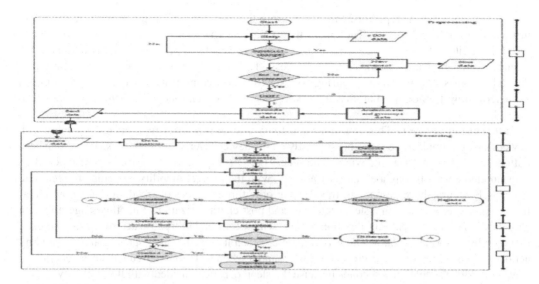

FIGURE 9.6   Processing stages of LoMoCA.

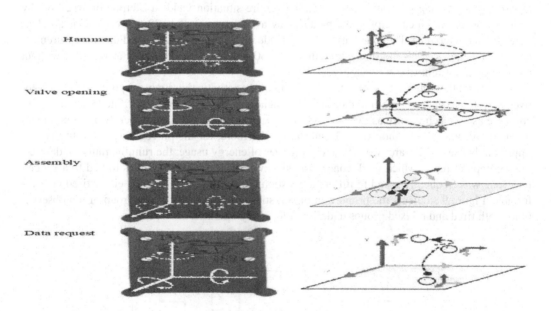

FIGURE 9.7   Movement patterns for experimentation.

LoMoCA Experiments: To test as many of the variations as possible, then assess the scheme, this chapter defines five baseline motions relevant to economic activities. In each case, a diagram with related angles as well as a trajectory interpretation is shown in Figure 9.7. The so-called lever, valve activation, and assembling motions (motions considered to reflect standard manufacturing activities for checking algorithms with such a progressive complexity rise) were categorized into two parts: the first one involves information demand and notifications, and the other includes data demand and alerts. To keep the number of stimuli minimal and to impede the character development process, both gestures were registered as shorter gestures. Here is a more in-depth description:

The 1–2–3 series was replicated three times with the hammer, which mostly uses two axes and just a single rotating direction. The OperaBLE XY plane was used to conduct this movement. It faced an additional challenge as a result of centrifugal acceleration, which causes a Z-axis difference, which obstructs the recognition process. The 1–2–3 series was performed four times. Assembly consists of link tasks requiring six axes (the most complicated motion), which is useful for determining LoMoCA reliability and distinguishing between identical gestures. It's 1–2–3 in the right order.

Information request: This is a unique activity that allows operations to test samples with devices using a simple three-tap process mostly on the device layer. It does not make use of displays, buttons, or other complicated devices. The series 1–2–3 is replicated three times. The OperaBLE consumption curves throughout hammer activity, with the phases, mobility, and idle are recorded. OperaBLE accurately represents portrayal using only three degrees of freedom (DOF) (acceleration data). It was also kept in a sleep state pre- and post-characterizations (mobility stage) to aid in classification of the various consumption phases, with a 1-second idle phase afterward to guarantee that mobility information gathering was complete. When it came to valve opening and assembling actions, the spending time on each action was around 3 seconds. In these circumstances, OperaBLE stayed in a sleep state after 2 seconds because the sleep period coincided with flow velocity.

The amount of time spent characterizing the movements is an important issue. OperaBLE utilizes a maximum of 0.011 mAh during the 6-second measuring interval set in this research (assembly movement). This suggests that prototypes will run for a minimum of 15.70 hours with six DOF utilizing this working process (worst case). The entire situation yields a lifetime of 10.27 hrs by splitting the 105 mAh capacity of the batteries by average usage of 10.22 mAh (motions are consistently characterized). The longevity attained is ideal, even though it is based upon measurement data. Even if the device's lifetime were reduced by 30%, it will still last a regular workday without any energy constraints.

Aside from the time being spent characterizing every activity, this study proposes a set of objectives as well as a decision-making mechanism depending on the number of DOF to be examined to conserve as so much energy as feasible. As mentioned in earlier sections, OperaBLE's first objective is stability to satisfy Industry 4.0 criteria. Furthermore, for the procedures, durability is also important because there are two primary categories of energy usage: the running rate of a device's microcomputer (upon which work concentrates) as well as the sensor modules linked to a gadget. Everyone was frequently asked to fill out a questionnaire stating that they felt either tired or comfortable. Figure 9.8 depicts the proportion of statistics correctness for every group of ten observations, with tired and relaxed groups underlined for every operator.

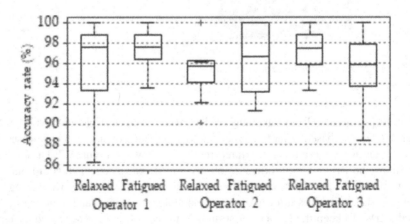

FIGURE 9.8   Statistical accuracy rate (%).

9.16 CONCLUSION

This chapter addresses how wearable devices can be used for tailored safety management surveillance and trends. A thorough examination of the characteristics of wearable devices is offered, as well as the associated safety metrics that are assumed to be effective in forecasting safety quality and leadership strategies. According to the study, wearable technologies are utilized in a range of industries to improve safety and productivity, but just a few applications have been found in building in Industry 4.0.

Given that the many industries where wearable techniques have been widely implemented are not growth industries, such as building, there seems to be an immediate need to preserve the status quo with respect to wearable application in building projects. It's past time for construction firms and professionals to adopt these new and emerging technologies to significantly improve safety management. This chapter identified wearable tech possibilities for gathering and tracking several measures linked to frequent building site accidents and deaths. According to the findings of this chapter, the detectors and major mechanisms in existent wearable devices used for other industries can indeed be designed to measure and analyze a wide range of safety performance measures in building.

Per the results of this study, some of the devices and software utilized in commonly available smart technology possess weaknesses and strengths that can be effectively handled by combining two or even more detecting processes to create reward systems. Wearables with numerous sensor types, as well as multi-modal sensors that may gather information from a data area of the body, were also proposed for use in building. Device manufacturers should focus their efforts on figuring out how to extract meaning from many sensors integrated into portable sensors to provide a holistic perspective as to how the program works or performs across many sensors and devices.

This finding of the study can even be utilized to incorporate various wearable sensors and technologies into construction-specific smart technology. Models for individual safety monitoring could be evaluated. Future research on the topic may include electing potential wearable technology from the commonly available ones that could be used in industry or building models for construction-specific wearables depending on the results of this study and conducting an experimental test to confirm their efficacy.

A modification of the prototype is in the works to reduce its size and improve simple patterns. More techniques are constantly being designed to improve OperaBLE's versatility and monitor the environs of operators, guaranteeing worker safety. BLE mesh networks are currently being constructed for assessing a comprehensive BLE-based CPS because BLE technology enhances flexibility and opens up opportunities for complex datasets. The findings are encouraging, indicating that more research is needed to assess durability and improved productivity in real-world Industry 4.0 scenarios.

REFERENCES

[1] Demirkesen, S., & Arditi, D. (2015). Construction safety personnel's perceptions of safety training practices. *International Journal of Project Management*, 33(5), 1160–1169.

[2] Hallowell, M. R. (2012). Safety-knowledge management in American construction organizations. *Journal of Management in Engineering*, 28(2), 203–211.

[3] Cheng, T., Migliaccio, G. C., Teizer, J., & Gatti, U. C. (2013). Data fusion of real-time location sensing and physiological status monitoring for ergonomics analysis of construction workers. *Journal of Computing in Civil Engineering*, 27(3), 320–335.

[4] Ewing-Nelson, C. (2021). *Another 275,000 Women Left the Labor Force in January*. National Women's Law Center.

[5] Liu, D., Jin, Z., & Gambatese, J. (2020). Scenarios for integrating IPS–IMU system with BIM technology in construction safety control. *Practice Periodical on Structural Design and Construction*, 25(1), 05019007.

[6]   Li, Q., Brannen, L., Rasoulkhani, K., Mostafavi, A., Stoa, R., Chowdhury, S., . . . & Jaselskis, E. (2020).
       Regulatory adaptation in the construction industry: Case study of the OSHA update to the respirable crys-
       talline silica standard. *Journal of Legal Affairs and Dispute Resolution in Engineering and Construction*,
       12(4), 06520003.

[7]   Ahmed, S. M., Azhar, S., & Forbes, L. H. (2006). Costs of injuries/illnesses and fatalities in construction
       and their impact on the construction economy. In *Proc., Int. Conf. on Global Unity for Safety and Health
       in Construction* (pp. 363–371).

[8]   ILO (International Labour Organization). (2017). *Conducting Labor Inspections on Construction: A
       Guide for Labor Inspectors.* Accessed January 15, 2019.

[9]   Dai, J., Goodrum, P. M., & Maloney, W. F. (2009). Construction craft workers' perceptions of the factors
       affecting their productivity. *Journal of Construction Engineering and Management,* 135(3), 217–226.

[10]   Taylor, T. R., Uddin, M., Goodrum, P. M., McCoy, A., & Shan, Y. (2012). Change orders and lessons
       learned: Knowledge from statistical analyses of engineering change orders on Kentucky highway proj-
       ects. *Journal of Construction Engineering and Management,* 138(12), 1360–1369.

[11]   Shan, Y., Zhai, D., Goodrum, P. M., Haas, C. T., & Caldas, C. H. (2016). Statistical analysis of the
       effectiveness of management programs in improving construction labor productivity on large industrial
       projects. *Journal of Management in Engineering,* 32(1), 04015018.

[12]   Ranasinghe, U., Ruwanpura, J., & Liu, X. (2012). Streamlining the construction productivity improve-
       ment process with the proposed role of a construction productivity improvement officer. *Journal of
       Construction Engineering and Management,* 138(6), 697–706.

[13]   Mohammadi, A., Tavakolan, M., & Khosravi, Y. (2018). Factors influencing safety performance on con-
       struction projects: A review. *Safety Science,* 109, 382–397.

[14]   Mitropoulos, P., Abdelhamid, T. S., & Howell, G. A. (2005). A systems model of construction accident
       causation. *Journal of Construction Engineering and Management,* 131(7), 816–825.

[15]   Hwang, S., & Lee, S. (2017). Wristband-type wearable health devices to measure construction workers'
       physical demands. *Automation in Construction,* 83, 330–340.

[16]   Ghodrati, N., Yiu, T. W., & Wilkinson, S. (2018). Unintended consequences of management strategies for
       improving labor productivity in the construction industry. *Journal of Safety Research,* 67, 107–116.

[17]   Karimi, H., Taylor, T. R., & Goodrum, P. M. (2017). Analysis of the impact of craft labor availability on
       North American construction project productivity and schedule performance. *Construction Management
       and Economics,* 35(6), 368–380.

[18]   McKinsey & Company. 2015. *The Construction Productivity Imperative.* Accessed January 15, 2018.
       www.mckinsey.com/industries/capital-projects-and-infrastructure/our-insights/the-construction-
       productivity-imperative.

[19]   Lu, Y. (2017). Industry 4.0: A survey on technologies, applications, and open research issues. *Journal of
       Industrial Information Integration,* 6, 1–10.

[20]   Borgia, E. (2014). The Internet of Things vision: Key features, applications, and open issues. *Computer
       Communications,* 54, 1–31.

[21]   Rajkumar, R., Lee, I., Sha, L., & Stankovic, J. (2010, June). Cyber-physical systems: The next computing
       revolution. In *Design Automation Conference* (pp. 731–736). IEEE.

[22]   Shhedi, Z. A., Moldoveanu, A., Moldoveanu, F., & Taslitchi, C. (2015, November). Real-time hand
       hygiene monitoring system for HAI prevention. In *2015 E-Health and Bioengineering Conference
       (EHB)* (pp. 1–4). IEEE.

[23]   Cai, F., Yi, C., Liu, S., Wang, Y., Liu, L., Liu, X., . . . & Wang, L. (2016). Ultrasensitive, passive, and
       wearable sensors for monitoring human muscle motion and physiological signals. *Biosensors and
       Bioelectronics,* 77, 907–913.

[24]   Van Lummel, R. C., Ainsworth, E., Lindemann, U., Zijlstra, W., Chiari, L., Van Campen, P., & Hausdorff,
       J. M. (2013). Automated approach for quantifying the repeated sit-to-stand using one body-fixed sensor
       in young and older adults. *Gait & Posture,* 38(1), 153–156.

[25]   Urmat, S., & Yalçın, M. E. (2015, November). Design and implementation of an ARM-based embedded
       system for pedestrian dead reckoning. In *2015 9th International Conference on Electrical and Electronics
       Engineering (ELECO)* (pp. 885–889). IEEE.

[26]   Lee, W., Seto, E., Lin, K. Y., & Migliaccio, G. C. (2017). An evaluation of wearable sensors and
       their placements for analyzing construction worker's trunk posture in laboratory conditions. *Applied
       Ergonomics,* 65, 424–436.

[27] Lee, W., Lin, K. Y., Seto, E., & Migliaccio, G. C. (2017). Wearable sensors for monitoring on-duty and off-duty worker physiological status and activities in construction. *Automation in Construction*, 83, 341–353.

[28] Chen, J., Qiu, J., & Ahn, C. (2017). Construction worker's awkward posture recognition through supervised motion tensor decomposition. *Automation in Construction*, 77, 67–81.

[29] Akhavian, R., & Behzadan, A. H. (2016). Smartphone-based construction workers' activity recognition and classification. *Automation in Construction*, 71, 198–209.

[30] Oliver, M., Schofield, G. M., Badland, H. M., & Shepherd, J. (2010). Utility of accelerometer thresholds for classifying sitting-in office workers. *Preventive Medicine*, 51(5), 357–360.

[31] Sheehan, K. J., Greene, B. R., Cunningham, C., Crosby, L., & Kenny, R. A. (2014). Early identification of declining balance in higher functioning older adults, an inertial sensor-based method. *Gait & Posture*, 39(4), 1034–1039.

[32] Park, J. S., Robinovitch, S., & Kim, W. S. (2015). A wireless wristband accelerometer for monitoring of rubber band exercises. *IEEE Sensors Journal*, 16(5), 1143–1150.

[33] Hallowell, M. R., Hinze, J. W., Baud, K. C., & Wehle, A. (2013). Proactive construction safety control: Measuring, monitoring, and responding to safety leading indicators. *Journal of Construction Engineering and Management*, 139(10), 04013010.

[34] Hinze, J., Thurman, S., & Wehle, A. (2013). Leading indicators of construction safety performance. *Safety Science*, 51(1), 23–28.

[35] Park, J., Kim, K., & Cho, Y. K. (2017). The framework of automated construction-safety monitoring using cloud-enabled BIM and BLE mobile tracking sensors. *Journal of Construction Engineering and Management*, 143(2), 05016019.

[36] Rebolj, D., Babič, N. Č., Magdič, A., Podbreznik, P., & Pšunder, M. (2008). Automated construction activity monitoring system. *Advanced Engineering Informatics*, 22(4), 493–503.

[37] Navon, R., & Sacks, R. (2007). Assessing research issues in automated project performance control (APPC). *Automation in Construction*, 16(4), 474–484.

[38] Ferreira, J. J., Fernandes, C. I., Rammal, H. G., & Veiga, P. M. (2021). Wearable technology and consumer interaction: A systematic review and research agenda. *Computers in Human Behavior*, 106710.

[39] Salahuddin, M., & Romeo, L. (2020). Wearable technology: Are product developers meeting consumer's needs? *International Journal of Fashion Design, Technology, and Education*, 13(1), 58–67.

[40] Krey, M. (2020, March). Wearable technology in health care—Acceptance and technical requirements for medical information systems. In *2020 6th International Conference on Information Management (ICIM)* (pp. 274–283). IEEE.

[41] Erkiliç, C. E., & Yalçin, A. (2020). Evaluation of the wearable technology market within the scope of digital health technologies. *Gazi İktisat ve İşletme Dergisi*, 6(3), 310–323.

[42] Krey, M., Schlatter, U., Mahadevan, D. A., Derungs, K., & Oehler, J. K. (2020). Wearable technology in healthcare. *Journal of Advances in Information Technology*, 11(3), 172–180.

[43] Almusawi, H. A., Durugbo, C. M., & Bugawa, A. M. (2021). Innovation in physical education: Teachers' perspectives on readiness for wearable technology integration. *Computers & Education*, 167, 104185.

[44] Awolusi, I., Nnaji, C., & Okpala, I. (2020, November). Success factors for the implementation of wearable sensing devices for safety and health monitoring in construction. In *Construction Research Congress 2020: Computer Applications* (pp. 1213–1222). Reston, VA: American Society of Civil Engineers.

[45] Zhang, M., Saeed, R., Stankovski, S., Xiaoshuan, Z., Saeed, S., & Zhang, X. (2020). Wearable technology and applications: A systematic review. *Journal of Mechatronics, Automation and Identification Technology*, 5, 5–16.

[46] Okpala, I., Parajuli, A., Nnaji, C., & Awolusi, I. (2020, November). Assessing the feasibility of integrating the Internet of Things into safety management systems: A focus on wearable sensing devices. In *Construction Research Congress 2020: Computer Applications* (pp. 236–245). Reston, VA: American Society of Civil Engineers.

[47] Salahuddin, M., & Lee, Y. A. (2020). Identifying key quality features for wearable technology embedded products using the Kano model. *International Journal of Clothing Science and Technology*, 33(1), 93–105. https://doi.org/10.1108/IJCST-08-2019-0130

[48] Morcos, M. W., Teeter, M. G., Somerville, L. E., & Lanting, B. (2020). Correlation between hip osteoarthritis and the level of physical activity as measured by wearable technology and patient-reported questionnaires. *Journal of Orthopedics*, 20, 236–239.

[49] Blount, D. S., McDonough, D. J., & Gao, Z. (2021). Effect of wearable technology-based physical activity interventions on breast cancer survivors' physiological, cognitive, and emotional outcomes: A systematic review. *Journal of Clinical Medicine*, 10(9), 2015.

[50] Erkiliç, C. E., & Yalçin, A. (2020). Evaluation of the wearable technology market within the scope of digital health technologies. *Gazi İktisat ve İşletme Dergisi*, 6(3), 310–323.

10 Analysis of Factors Influencing Cloud Computing Adoption in Industry 4.0-Based Advanced Manufacturing Systems

Anilkumar Malaga and S. Vinodh

CONTENTS

10.1 Introduction..91
10.2 Literature Review...92
10.3 Significance of Cloud Computing 93
    10.3.1 Cloud Computing in Industry 4.0..93
    10.3.2 Cloud Manufacturing...93
10.4 Case Study 94
10.5 Conclusion 100
References..101

10.1 INTRODUCTION

Innovation is the outcome of transformation in the competitiveness of companies that can develop or reach new markets [1]. A new era of manufacturing called Industry 4.0 (I4.0), through new approaches, evolved to manage manufacturing with the progress of computer and electro-mechanical technologies [2]. Cloud computing (CC) facilitates flexible and dynamic infrastructure, assured service quality, and adaptable software services. Normally, CC is a commercial objective for the creation of a field network revolution. It extends and develops simultaneous, decentralized, and grid computing services. Cloud computing provides various advantages, including infrastructure and operational cost reduction [3]. As a new technology, CC could contribute to improved cost effectiveness and timely access to services [4]. The cloud computing paradigm is a way to analyze a computer pool on-demand network [5]. For businesses that have restricted financial capacities, cloud computing is developed as an option. It has become a necessity that companies employ technology. However, companies must have the necessary resources such as IT infrastructure, software, and hardware, or they would not have the capacity to do so [6].

Due to the enormous scale of cloud technology and the intricacy of its service relationships, cloud computing has numerous definitions. The US National Institute of Standards and Technology (NIST) provided a widely cited definition: "Cloud computing is characterized as a concept enabling convenient, on-demand network access to a joint pool of programmable computing resources, which can be provided promptly and released with minimum administrative effort or interconnection among services providers" [6]. Cloud computing has transformed computer services from an investment in infrastructure to services that may be accessed anytime and anywhere. A growing need for rapid service delivery was requested for cloud computing to promote IT agility by companies [4]. General IT strategies of large and small organizations are quickly being redirected to cloud computing. Although this new technology presents advantages, there is evidence that not all enterprises are eager to employ cloud-based solutions [7]. Businesses with minimal capacity

DOI: 10.1201/9781003186670-10

have not been able to upgrade technology at their own expense. However, through the development of cloud computing and a broad range of providers, all enterprises have been able to utilize cloud computing. One of the primary advantages of cloud-based services is cost reduction. Savings are not just in money but in time and effort to manage and run the system [6]. Cloud computing services have already been considered one of the best world's leading business strategies, and the influential factors need to be analyzed. This study focuses on identifying the influential factors of CC adoption in Industry 4.0–based manufacturing systems. After identification, the factors are prioritized using a fuzzy Technique for Order Preference by Similarity to Ideal Solution (TOPSIS) approach.

The rest of this chapter is organized as follows: Section 10.2 analyzes the literature, Section 10.3 discusses the significance of cloud computing in the industry, Section 10.4 presents the case study, and Section 10.5 details the conclusions of this study.

10.2   LITERATURE REVIEW

Oliveira et al. [7] identified cloud adoption drivers depending on innovative characteristics, technology, organization, and environmental circumstances. The authors examined cloud adoption determinants in the production and service industries. The outcome of this work presents that a company's use of cloud computing has a direct influence on its relative benefit, complexity, technical preparedness, top management support, and organization size. Results assessment validated the direct impact of cost savings on cloud computing's relative benefit and its indirect influence on cloud adoption.

Hasan et al. [6] examined the literature to identify factors that impact the implementation of cloud computing from existing research. The authors developed a conceptual model through the retrieval of factors that impact the adoption of CC. The authors revealed that the elements that drive CC acceptance include perceived simplicity to use, perceived value, security, compatibility, costs, and management assistance. For future work, the authors recommended qualitative investigation and empirical examination of the presented model.

Johnston et al. [8] offered a picture of the commercial value of cloud computing using qualitative information through an explorative study. The outcomes of the assessment serve to create a theoretical framework. This work contributes to the knowledge gathered via an understanding of cloud computing's business value in South Africa. The authors concluded that the investigation facilitated insights into the aspects that impact cloud computing business value in South Africa. It provides awareness of the future for research and development for businesses and academia.

Jayasekara et al. [9] developed a framework to evaluate industry preparedness for cloud manufacturing. The authors identified elements that influence cloud manufacturing readiness, essential features, essential business processes, and production needs for transforming conventional production organizations. The authors mentioned that this work facilitates decision-makers in the production sector to grasp the prerequisites required for effective implementation of cloud manufacturing.

Liu et al. [10] examined significant challenges with cloud production and outlined their prospects for the future. The idea, operational model, service model, service content, technological system, architecture, and key features of cloud production were addressed. The conversation was conducted in comparison with challenges in cloud computing, which helps to understand the difficulties of cloud construction better. The authors also explored the interconnections among cloud production and several other production-related ideas, including cloud computing, CPS, intelligent production, Industry 4.0, and Industrial Internet.

Misra et al. [11] recognized variables contributing to technology, organization, and environment to implement cloud services in the semiconductor production industry effectively. Regression analysis was used to evaluate the surveyed information. The authors revealed that product and service market globalization played a significant part in delivering industry development, and the exchange of information has been made simpler by globalization. The authors concluded that integrating the identified successful variables into strategic and operational planning can minimize risk and facilitate the adoption of semi-manufacturing cloud technologies.

Vater et al. [5] proposed constructing a prescribed automation IT design depending on edge and cloud computing. The proposed design allows interoperable process control depending on the network. It gives the capability of extensive data processing to boost the productivity of the production process continually. The design provided the potential to regularly enhance productivity in the production process through comprehensive data management. The authors concluded that edge and cloud computing were the major processing elements in this system.

Alsafi and Fan [12] facilitated an update on the assessment of main obstacles at the present level of cloud computing faced by SMEs in Saudi Arabia. The most prevalent impediments were identified. The authors stated that SMEs might deal with challenges and use cloud computing in their companies if appropriately prepared. The authors concluded that overcoming the obstacles mentioned could enable the Saudi government and small and medium-sized businesses to establish effective cloud computing migration plans.

10.3 SIGNIFICANCE OF CLOUD COMPUTING

10.3.1 CLOUD COMPUTING IN INDUSTRY 4.0

The advent of the Industrial Revolution was because of the progress of non–human-driven equipment. The concept of mass manufacture of products and services was introduced in all sectors. Later on, computer-controlled machinery was designed to improve efficiency even more with the growth of computer technology. The focus of manufacturing is shifting towards an interdisciplinary perspective. Various areas of science and engineering were devised to assist operators in working effectively. As technology has evolved in multiple disciplines, this has also led to industry expansion and vice versa. The industry is on the edge of a new revolution. The new paradigm generation is termed Industry 4.0. In Industry 4.0, new approaches to manufacturing have emerged with the growth of computer, electrical, and mechanical technologies. In the previous scenario, only networking could be utilized through a local area network. But now, cloud computing enables controlling or monitoring of resources through the Internet. Cloud computing incorporates the service-oriented concept. CC is categorized into different models [2]. The pay-per-view access to cloud services is a less expensive replacement for traditional data centers. Clouds often facilitate on-demand services and network data center virtualization, which is an expected solution to address performance, flexibility, reduced cost, scalability, improvement in resource usage, and energy efficiency [3]. Cloud technology can respond if a firm has its own infrastructure, the problem of business instabilities, and changes in the requirements of resources and services. It is a strategic choice for an organization to migrate to the cloud. However, certain factors, which are detailed in this chapter, might affect this decision. A variety of advantages are offered by cloud technology, whose effective implementation and daily use have several positive consequences in the areas of company improvement, saving costs, and improving productive activities.

10.3.2 CLOUD MANUFACTURING

Cloud manufacturing is the modern paradigm. Cloud manufacturing is considered a new multidisciplinary area that utilizes and includes intelligent technologies such as network-based production, virtual production, agile manufacturing, and the manufacturing grid, as well as cloud computing that reflects both "distributed resources integration" and "integrating resources distribution" [9]. Cloud manufacturing adopts the concept of "cloud computing," which includes the idea of Manufacturing-as-a-Service (MaaS); that is, all production resources and abilities included in the lifetime of the product are encased in services. MaaS also has three variants in service models: Manufacturing Platform-as-a-Service (MPaaS), Manufacturing Infrastructure-as-a-Service

(MIaaS), and Manufacturing Software-as-a-Service (MSaaS) [10]. The idea of "manufacturing" should be generally viewed from the viewpoint of cloud manufacturing.

Big corporations such as Amazon, Google, Facebook, and Yahoo are now using data centers on a daily basis to store, search the web, and compute on a vast scale. Hosting data centers has become a huge investment with the introduction of cloud computing services, which is a key factor in the future of IT organizations [3].

10.4 CASE STUDY

The goal of this chapter is to evaluate the most influential factors using a suitable approach for cloud computing adoption. Several factors are identified that impact CC adoption in I4.0-based modern manufacturing systems. Appropriate criteria have been defined, and main factors can be evaluated in a fuzzy environment using multi-criteria decision-making approach TOPSIS. The inputs were obtained, and necessary calculations were carried out to assess the ranking of factors by experts.

The case study was carried out for a manufacturing organization producing automobile components. Industry 4.0 is being implemented by the organization. This chapter examines the elements that influence cloud concept adoption of I4.0 production system implementation using a fuzzy TOPSIS approach.

The detailed procedure of evaluation using fuzzy TOPSIS in this chapter is as follows [13]:

Step 1: Identify criteria and factors influencing CC adoption in I4.0-based production systems through literature and industries implementing CC and Industry 4.0.

The factors influencing cloud computing adoption of I4.0-based advanced manufacturing systems are identified from the literature.

Step 2: Identify appropriate members for the expert panel from academia and industries.

An expert panel was formed consisting of experts from academia, research and development, and industry to evaluate and provide inputs. The identified factors have been validated by experts and are depicted in Table 10.1.

Identified eighteen factors were evaluated under four criteria: Technology (C1), Organization (C2), Environment (C3), and Economy (C4). Eighteen factors under four criteria were presented to the expert panel in order to collect inputs in linguistic terms on a 7-point scale.

Step3: Recognize the suitable linguistic terms to collect inputs for criteria weights and factor ratings from expert panel.

The linguistic representations of the factor ratings were transformed into fuzzy numbers using the scale: Very poor (VP)—(0, 0, 1, 2), Poor (P)—(1, 2, 2, 3), Medium Poor (MP)—(2, 3, 4, 5), Fair (F)—(4, 5, 5, 6), Medium Good (MG)—(5, 6, 7, 8), Good (G)—(7, 8, 8, 9), Very Good (VG)—(8, 9, 10, 10). The scale used for criteria weights was: Very Low (VL)—(0, 0, 0.1, 0.2), Low (L)—(0.1, 0.2, 0.2, 0.3), Medium Low (ML)—(0.2, 0.3, 0.4, 0.5), Medium (M)—(0.4, 0.5, 0.5, 0.6), Medium High (MH)—(0.5, 0.6, 0.7, 0.8), High (H)—(0.7, 0.8, 0.8, 0.9), Very High (VH)—(0.8, 0.9, 1, 1).

Step 4: Collection of inputs from experts for factor ratings and criteria weights.

Factor ratings and criteria weights were collected from the expert panel and are depicted in Table 10.2.

TABLE 10.1

Factors Influencing CC Adoption in I4.0-Based Advanced Manufacturing Systems

Factor	Research Study
Compatibility (F1)	Oliveira et al. [7], Badie et al. [3], Hasan et al. [6], Yoo and Kim [14]
Competitive pressure (F2)	Oliveira et al. [7], Badie et al. [3], Hasan et al. [6], Misra et al. [11], Yoo and Kim [14]
Complexity (F3)	Oliveira et al. [7], Badie et al. [3], Hasan et al. [6]
Computational efficiency (F4)	Misra et al. [11]
Cost (F5)	Oliveira et al. [7], Hasan et al. [6], Ali et al. [4], Yoo and Kim [14]
Customization (F6)	Badie et al. [3], Misra et al. [11], Hasan et al. [6]
Data backup (F7)	Ali et al. [4], Vater et al. [5]
Firm size (F8)	Oliveira et al. [7], Badie et al. [3], Hasan et al. [6], Misra et al. [11]
Government regulation (F9)	Yoo and Kim [14], Misra et al. [11], Oliveira et al. [7]
Innovation (F10)	Hasan et al. [6]
Internet connectivity (F11)	Ali et al. [4], Dobrescu et al. [15]
Interoperability (F12)	Hasan et al. [6]
On-demand product and service availability (F13)	Misra et al. [11], Ali et al. [4]
Organization readiness (F14)	Yoo and Kim [14], Misra et al. [11]
Scalability (F15)	Misra et al. [11]
Security (F16)	Oliveira et al. [7], Badie et al. [3], Hasan et al. [6], Ali et al. [4]
Technology support infrastructure (F17)	Yoo and Kim [14], Wang et al. [16]
Top management support (F18)	Oliveira et al. [7], Badie et al. [3], Hasan et al. [6], Yoo and Kim [14], Misra et al. [11]

TABLE 10.2

Inputs of Factor Ratings and Criteria Weights Provided by Expert Panel

Factors	Technology (C1)			Organization (C2)			Environment (C3)			Economy (C4)		
Experts	D1	D2	D3	D1	D2	D3	D1	D2	D3	D1	D2	D3
F1	VG	G	VG	G	MG	G	G	F	G	V	G	VG
F2	G	MG	G	VG	G	G	VG	G	VG	G	F	G
F3	VG	G	VG	G	G	G	G	F	G	G	MG	G
F4	G	MG	G	G	VG	G	VG	F	VG	VG	G	VG
F5	F	MG	F	F	MG	MP	G	MG	G	G	F	G
F6	G	MG	G	G	MP	G	VG	G	VG	VG	G	G
F7	G	MG	G	G	F	MG	G	MG	G	G	F	G
F8	VG	G	G	G	MG	G	F	MG	MP	G	MG	G
F9	G	VG	G	VG	G	G	G	MG	G	G	F	G
F10	G	MG	G	G	MG	G	VG	G	VG	VG	G	VG
F11	VG	G	VG	G	VG	G	G	MG	G	G	F	G
F12	G	MG	G	VG	G	G	G	F	G	VG	G	VG
F13	MG	F	MP	G	MG	G	VG	G	VG	G	F	G
F14	G	MG	G	G	F	MG	G	MG	G	G	F	G
F15	G	VG	G	VG	G	G	G	VG	G	G	VG	G
F16	G	MG	G	G	MG	G	G	F	G	VG	G	VG
F17	VG	G	G	G	MG	G	G	MP	G	G	F	G
F18	G	MG	G	VG	G	VG	G	F	G	G	F	G
Experts	**Technology (C1)**			**Organization (C2)**			**Environment (C3)**			**Economy (C4)**		
DM1	VH			H			MH			H		
DM2	H			MH			H			V		
DM3	V			H			M			H		

Step 5: Aggregation of factor ratings and criteria weights using Equations 1 and 2.

$$\text{Aggregation of factor ratings } A_{ij} = (A_{ij}^1, A_{ij}^x, A_{ij}^y, A_{ij}^z) \tag{1}$$

$$\text{Aggregation of criteria weights } W_j = (W_j^1, W_j^x, W_j^y, W_j^z) \tag{2}$$

Where

$$A_{ij}^1 = \text{Min}_e (A_{ije}^1); A_{ij}^x = \frac{1}{e}\sum_{e=1}^{E}(A; A_{ij}^y = \frac{1}{e}\sum_{e=1}^{E}(A; A_{ij}^z = \text{Max}_e (A_{ije}^z)$$

$$W_j^1 = \text{Min}_e (W_{je}^1); W_j^x = \frac{1}{e}\sum_{e=1}^{E}(w; W_j^y = \frac{1}{e}\sum_{e=1}^{E}(w; W_j^z = \text{Max}_e (W_{je}^z)$$

The collected inputs in linguistic terms were converted into fuzzy numbers. Then the fuzzy numbers were aggregated using Equations 1 and 2.

Sample calculation:

For factor F1 under criteria C1, the inputs from three experts were VG, G, and VG. In fuzzy numbers, the inputs were (8,9,10,10); (7,8,8,9), and (8,9,10,10).

Aggregation of F1 using Equation 1:

$$A_{11}^1 = \text{Min}(8,7,8) = 7, A_{11}^2 = \frac{1}{3}\left(\text{Sum}(9,8,9)\right) = 8.67, A_{11}^3 = \frac{1}{3}\left(\text{Sum}(10,8,10)\right) = 9.33, \text{ and } A_{11}^4$$
$$= \text{Max}(10,9,10) = 10.$$

Similarly, for the criteria weight of C1, the inputs from three experts were VH, H, and VH. In fuzzy numbers, the weights were (0.8,0.9,1,1); (0.7,0.8,0.8,0.9), and (0.8,0.9,1,1).

Aggregation of weight for C1 using Equation 2:

$$W_1^1 = \text{Min}(0.8,0.7,0.8) = 0.7; W_1^2 = \frac{1}{3}\left(\text{Sum}(0.9,0.8,0.9)\right) = 0.87, W_1^3 = \frac{1}{3}\left(\text{Sum}(1,0.8,1)\right)$$
$$= 0.93, \text{ and } W_1^4 = \text{Max}(1,0.9,1) = 1$$

The aggregation for each factor rating and criteria weight was calculated using Equations 1 and 2, and they are presented in Table 10.3.

Step 6: Normalization of the decision matrix.

Then the decision matrix was normalized. Hence, the normalized decision matrix $N = [N_{ij}]_{mx}n$ [17].

$$\text{For beneficial criteria } [N_{ij}] = (N_{ij}^1, N_{ij}^x, N_{ij}^y, N_{ij}^z) = (A_{ij}^1/A_j^{z+}, A_{ij}^x/A_j^{z+}, A_{ij}^y/A_j^{z+}, A_{ij}^z/A_j^{z+}) \tag{3}$$

$$\text{For non-beneficial criteria } [N_{ij}] = (A_j^{1-}/A_{ij}^z, A_j^{1-}/A_{ij}^y, A_j^{1-}/A_{ij}^x, A_j^{1-}/A_{ij}^1) \tag{4}$$

Where $A_j^{z+} = \text{Max}_e(A_{ij}^z)$
$\qquad A_j^{1-} = \text{Min}_e(A_{ij}^1)$

In this chapter, among four criteria, economy is considered a cost criterion, that is, non-beneficial, and the remining three are considered beneficial criteria.

Hence, the normalization of the aggregated decision matrix was computed using Equations 3 and 4.

TABLE 10.3
Aggregation of Factor Ratings and Criteria Weights

Factors	C1	C2	C3	C4
F1	(7,8.67,9.33,10)	(5,7.33,7.67,9)	(4,7,7,9)	(7,8.67,9.33,10)
F2	(5,7.33,7.67,9)	(7,8.33,8.67,10)	(7,8.67,9.33,10)	(4,7,7,9)
F3	(7,8.67,9.33,10)	(5,7.33,7.67,9)	(4,7,7,9)	(5,7.33,7.67,9)
F4	(5,7.33,7.67,9)	(7,8.33,8.67,10)	(4,7.67,8.33,10)	(7,8.67,9.33,10)
F5	(4,5.33,5.67,8)	(2,4.67,5.33,8)	(5,7.33,7.67,9)	(4,7,7,9)
F6	(5,7.33,7.67,9)	(2,6.33,6.67,9)	(7,8.67,9.33,10)	(7,8.33,8.67,10)
F7	(5,7.33,7.67,9)	(4,6.33,6.67,9)	(5,7.33,7.67,9)	(4,7,7,9)
F8	(7,8.33,8.67,10)	(5,7.33,7.67,9)	(2,4.67,5.33,8)	(5,7.33,7.67,9)
F9	(7,8.33,8.67,10)	(7,8.33,8.67,10)	(5,7.33,7.67,9)	(4,7,7,9)
F10	(5,7.33,7.67,9)	(5,7.33,7.67,9)	(7,8.67,9.33,10)	(7,8.67,9.33,10)
F11	(7,8.67,9.33,10)	(7,8.33,8.67,10)	(5,7.33,7.67,9)	(4,7,7,9)
F12	(5,7.33,7.67,9)	(7,8.33,8.67,10)	(4,7,7,9)	(7,8.67,9.33,10)
F13	(2,4.67,5.33,8)	(5,7.33,7.67,9)	(7,8.67,9.33,10)	(4,7,7,9)
F14	(5,7.33,7.67,9)	(4,6.33,6.67,9)	(5,7.33,7.67,9)	(4,7,7,9)
F15	(7,8.33,8.67,10)	(7,8.33,8.67,10)	(7,8.33,8.67,10)	(7,8.33,8.67,10)
F16	(5,7.33,7.67,9)	(5,7.33,7.67,9)	(4,7,7,9)	(7,8.67,9.33,10)
F17	(7,8.33,8.67,10)	(5,7.33,7.67,9)	(2,6.33,6.67,9)	(4,7,7,9)
F18	(5,7.33,7.67,9)	(7,8.67,9.33,10)	(4,7,7,9)	(4,7,7,9)
Weight	(0.7,0.87,0.93,1)	(0.5,0.73,0.77,0.9)	(0.4,0.63,0.67,0.9)	(0.7,0.83,0.87,1)

Sample calculation:

For factor F1 under C1: $A_j^{4+} = \text{Max}_i(A_{ij}^z) = \text{Max}(10,9,10,9,8,9,9,10,10,9,10,9,8,9,10,9,10,9) = 10$

normalization $[N]_{1\times1} = \left(\dfrac{7}{10}, \dfrac{8.67}{10}, \dfrac{9.33}{10}, \dfrac{10}{10} \right) = (0.7, 0.867, 0.933, 0.1)$

For factor F1 under C4, which is a non-beneficial criterion:

$A_j^{1-} = \text{Min}_i(A_{ij}^1) = \text{Min}\,(7, 4, 5, 7, 4, 7, 4, 5, 4, 7, 4, 7, 4, 4, 7, 7, 4, 4) = 4$

Normalization $[N]_{1\times4} = \left(\dfrac{4}{10}, \dfrac{4}{9.33}, \dfrac{4}{8.67}, \dfrac{4}{7} \right) = (0.4, 0.43, 0.46, 0.57)$

In the same way, the aggregated matrix is normalized using Equations 3 and 4, and the normalization matrix is depicted in Table 10.4.

Step 7: Computation of weighted normalized matrix.

In this step, the weighted matrix is taken as a product with normalized decision matrix.

The weighted normalized matrix: $V = [V_{ij}] = [W_j] \times [N_{ij}]_{m\times}n$ (5)

Sample calculation of weighted normalization for factor F1:

$[V_{11}] = (0.7, 0.87, 0.93, 1) \times (0.7, 0.87, 0.93, 1) = (0.49, 0.76, 0.86, 1)$

Then the weighted normalized matrix is computed using Equation 5, and the values are depicted in Table 10.5.

TABLE 10.4
Normalized Matrix

Factors	C1	C2	C3	C4
F1	(0.7,0.87,0.93,1)	(0.5,0.73,0.77,0.9)	(0.4,0.7,0.7,0.9)	(0.4,0.43,0.46,0.57)
F2	(0.5,0.73,0.77,0.9)	(0.7,0.83,0.87,1)	(0.7,0.87,0.93,1)	(0.44,0.57,0.57,1)
F3	(0.7,0.87,0.93,1)	(0.5,0.73,0.77,0.9)	(0.4,0.7,0.7,0.9)	(0.44,0.52,0.55,0.8)
F4	(0.5,0.73,0.77,0.9)	(0.7,0.83,0.87,1)	(0.4,0.77,0.83,1)	(0.4,0.43,0.46,0.57)
F5	(0.4,0.53,0.57,0.8)	(0.2,0.47,0.53,0.8)	(0.5,0.73,0.77,0.9)	(0.44,0.57,0.57,1)
F6	(0.5,0.73,0.77,0.9)	(0.2,0.63,0.67,0.9)	(0.7,0.87,0.93,1)	(0.4,0.46,0.48,0.57)
F7	(0.5,0.73,0.77,0.9)	(0.4,0.63,0.67,0.9)	(0.5,0.73,0.77,0.9)	(0.44,0.57,0.57,1)
F8	(0.7,0.83,0.87,1)	(0.5,0.73,0.77,0.9)	(0.2,0.47,0.53,0.8)	(0.44,0.52,0.55,0.8)
F9	(0.7,0.83,0.87,1)	(0.7,0.83,0.87,1)	(0.5,0.73,0.77,0.9)	(0.44,0.57,0.57,1)
F10	(0.5,0.73,0.77,0.9)	(0.5,0.73,0.77,0.9)	(0.7,0.87,0.93,1)	(0.4,0.43,0.46,0.57)
F11	(0.7,0.87,0.93,1)	(0.7,0.83,0.87,1)	(0.5,0.73,0.77,0.9)	(0.44,0.57,0.57,1)
F12	(0.5,0.73,0.77,0.9)	(0.7,0.83,0.87,1)	(0.4,0.7,0.7,0.9)	(0.4,0.43,0.46,0.57)
F13	(0.2,0.47,0.53,0.8)	(0.5,0.73,0.77,0.9)	(0.7,0.87,0.93,1)	(0.44,0.57,0.57,1)
F14	(0.5,0.73,0.77,0.9)	(0.4,0.63,0.67,0.9)	(0.5,0.73,0.77,0.9)	(0.44,0.57,0.57,1)
F15	(0.7,0.83,0.87,1)	(0.7,0.83,0.87,1)	(0.7,0.83,0.87,1)	(0.4,0.46,0.48,0.57)
F16	(0.5,0.73,0.77,0.9)	(0.5,0.73,0.77,0.9)	(0.4,0.7,0.7,0.9)	(0.4,0.43,0.46,0.57)
F17	(0.7,0.83,0.87,1)	(0.5,0.73,0.77,0.9)	(0.2,0.63,0.67,0.9)	(0.44,0.57,0.57,1)
F18	(0.5,0.73,0.77,0.9)	(0.7,0.87,0.93,1)	(0.4,0.7,0.7,0.9)	(0.44,0.57,0.57,1)

TABLE 10.5
Weighted Normalized Matrix

Factors	C1	C2	C3	C4
F1	(0.49,0.76,0.86,1)	(0.25,0.53,0.59,0.81)	(0.16,0.44,0.47,0.81)	(0.28,0.36,0.4,0.57)
F2	(0.35,0.64,0.72,0.9)	(0.35,0.61,0.67,0.9)	(0.28,0.55,0.62,0.9)	(0.31,0.47,0.5,1)
F3	(0.49,0.76,0.86,1)	(0.25,0.53,0.59,0.81)	(0.16,0.44,0.47,0.81)	(0.31,0.43,0.48,0.8)
F4	(0.35,0.64,0.72,0.9)	(0.35,0.61,0.67,0.9)	(0.16,0.49,0.56,0.9)	(0.28,0.36,0.4,0.57)
F5	(0.28,0.46,0.53,0.8)	(0.1,0.34,0.41,0.72)	(0.2,0.46,0.52,0.81)	(0.31,0.47,0.5,1)
F6	(0.35,0.64,0.72,0.9)	(0.1,0.46,0.52,0.81)	(0.28,0.55,0.62,0.9)	(0.28,0.38,0.42,0.57)
F7	(0.35,0.64,0.72,0.9)	(0.2,0.46,0.52,0.81)	(0.2,0.46,0.52,0.81)	(0.31,0.47,0.5,1)
F8	(0.49,0.72,0.81,1)	(0.25,0.53,0.59,0.81)	(0.08,0.3,0.36,0.72)	(0.31,0.43,0.48,0.8)
F9	(0.49,0.72,0.81,1)	(0.35,0.61,0.67,0.9)	(0.2,0.46,0.52,0.81)	(0.31,0.47,0.5,1)
F10	(0.35,0.64,0.72,0.9)	(0.25,0.53,0.59,0.81)	(0.28,0.55,0.62,0.9)	(0.28,0.36,0.4,0.57)
F11	(0.49,0.76,0.86,1)	(0.35,0.61,0.67,0.9)	(0.2,0.46,0.52,0.81)	(0.31,0.47,0.5,1)
F12	(0.35,0.64,0.72,0.9)	(0.35,0.61,0.67,0.9)	(0.16,0.44,0.47,0.81)	(0.28,0.36,0.4,0.57)
F13	(0.14,0.41,0.49,0.8)	(0.25,0.53,0.59,0.81)	(0.28,0.55,0.62,0.9)	(0.31,0.47,0.5,1)
F14	(0.35,0.64,0.72,0.9)	(0.2,0.46,0.52,0.81)	(0.2,0.46,0.52,0.81)	(0.31,0.47,0.5,1)
F15	(0.49,0.72,0.81,1)	(0.35,0.61,0.67,0.9)	(0.28,0.52,0.58,0.9)	(0.28,0.38,0.42,0.57)
F16	(0.35,0.64,0.72,0.9)	(0.25,0.53,0.59,0.81)	(0.16,0.44,0.47,0.81)	(0.28,0.36,0.4,0.57)
F17	(0.49,0.72,0.81,1)	(0.25,0.53,0.59,0.81)	(0.08,0.4,0.45,0.81)	(0.31,0.47,0.5,1)
F18	(0.35,0.64,0.72,0.9)	(0.35,0.64,0.72,0.9)	(0.16,0.44,0.47,0.81)	(0.31,0.47,0.5,1)

Step 8: Identification of ideal positive and negative solutions of individual criteria.

For beneficial criteria: Ideal positive solution (IPS) $I^+ = \text{Max}(V_{ij})$ (6)
Ideal negative solution (INS) $I^- = \text{Min}(V_{ij})$ (7)
For non-beneficial criteria: IPS $I^+ = \text{Min}(V_{ij})$ (8)
INS: $I^- = \text{Max}(V_{ij})$ (9)

Using Equations 6, 7, 8, and 9, the fuzzy ideal positive and negative solutions are calculated for both beneficial and non-beneficial criteria and are depicted in Table 10.6.

Step 9: Computation of separation of each factor from fuzzy ideal positive solution and ideal negative solution [17].

$$\text{Separation of each factor from IPS: } S_i^+ = \sqrt{\frac{1}{4}\sum_{i=1}^{n}\left(v_{ij} - I^+\right)^2} \tag{10}$$

$$\text{Separation of each factor from INS: } S_i^- = \sqrt{\sum_{i=1}^{n}\left(v_{ij} - I^-\right)^2} \tag{11}$$

Sample calculation:

Distance of factor F1 from ideal positive solution of C1 =

$$= \sqrt{\frac{1}{4}\left(0.49-0.49\right)^2 + \left(0.76-0.76\right)^2 + \left(0.86-0.86\right)^2 + \left(1-1\right)^2} = 0$$

Distance of factor F1 from ideal negative solution of C1 =

$$= \sqrt{\frac{1}{4}\left(0.49-0.14\right)^2 + \left(0.76-0.41\right)^2 + \left(0.86-0.49\right)^2 + \left(1-0.8\right)^2} = 0.3248$$

Similarly, the distance of all factors related to all criteria are computed and are depicted in Table 10.7.

Step 10: Computation of closeness coefficient of individual factor.

$$CC_i = \frac{S_i^-}{S_i^+ + S_i^-} \tag{12}$$

The closeness coefficients of individual factors are computed using Equation 12 and are depicted in Table 10.8.

Step 11: Finally, the order of preference of factors influencing CC adoption in Industry 4.0-based modern manufacturing systems has to be ranked. The prioritization of the factors influencing CC adoption is depicted in Table 10.8.

TABLE 10.6

Ideal Positive and Negative Solutions of Individual Criteria

Criteria	C1	C2	C3	C4
Ideal positive	(0.49,0.76,0.86,1)	(0.35,0.64,0.72,0.9)	(0.28,0.55,0.62,0.9)	(0.28,0.36,0.4,0.57)
Ideal negative	(0.14,0.41,0.49,0.8)	(0.1,0.34,0.41,0.72)	(0.08,0.3,0.36,0.72)	(0.31,0.47,0.5,1)

TABLE 10.7

Separation of Each Factor from Ideal Positive and Negative Solutions

	From ideal positive						From ideal negative				
	C1	C2	C3	C4	S_i^+		C1	C2	C3	C4	S_i^-
F1	0	0.1	0.1	0	0.228	F1	0.3	0.2	0.1	0.2	0.8176
F2	0.1	0	0	0	0.3832	F2	0.2	0.2	0.2	0	0.6675
F3	0	0.1	0.1	0	0.3556	F3	0.3	0.2	0.1	0.1	0.6921
F4	0.1	0	0.1	0	0.2287	F4	0.2	0.2	0.2	0.2	0.84
F5	0.3	0.3	0.1	0	0.8493	F5	0.1	0	0.1	0	0.2127
F6	0.1	0.2	0	0	0.3293	F6	0.2	0.1	0.2	0.2	0.7417
F7	0.1	0.2	0.1	0	0.6048	F7	0.2	0.1	0.1	0	0.4413
F8	0	0.1	0.2	0	0.4931	F8	0.3	0.2	0	0.1	0.5603
F9	0	0	0.1	0	0.3794	F9	0.3	0.2	0.1	0	0.6787
F10	0.1	0.1	0	0	0.2346	F10	0.2	0.2	0.2	0.2	0.8103
F11	0	0	0.1	0	0.3474	F11	0.3	0.2	0.1	0	0.7031
F12	0.1	0	0.1	0	0.2747	F12	0.2	0.2	0.1	0.2	0.778
F13	0.3	0.1	0	0	0.6613	F13	0	0.2	0.2	0	0.3824
F14	0.1	0.2	0.1	0	0.6048	F14	0.2	0.1	0.1	0	0.4413
F15	0	0	0	0	0.1003	F15	0.3	0.2	0.2	0.2	0.9725
F16	0.1	0.1	0.1	0	0.3541	F16	0.2	0.2	0.1	0.2	0.6928
F17	0	0.1	0.2	0	0.5262	F17	0.3	0.2	0.1	0	0.5387
F18	0.1	0	0.1	0	0.4736	F18	0.2	0.3	0.1	0	0.5725

TABLE 10.8

Ranking of the Factors Influencing CC Adoption

Factor	F1	F2	F3	F4	F5	F6	F7	F8	F9
S_i^+	0.23	0.383	0.356	0.229	0.849	0.329	0.605	0.493	0.379
S_i^-	0.82	0.668	0.692	0.84	0.213	0.742	0.442	0.56	0.679
CC_i	0.78	0.635	0.661	0.786	0.2	0.693	0.422	0.532	0.641
Rank	3	11	9	2	18	6	15	13	10

Factor	F10	F11	F12	F13	F14	F15	F16	F17	F18
S_i^+	0.23	0.347	0.275	0.661	0.605	0.1	0.354	0.526	0.474
S_i^-	0.81	0.703	0.778	0.382	0.441	0.973	0.693	0.539	0.573
CC_i	0.78	0.669	0.739	0.366	0.421	0.907	0.662	0.506	0.547
Rank	4	7	5	17	16	1	8	14	12

Table 10.8 represents the ranking of factors that influence CC adoption in Industry 4.0-based advanced manufacturing systems. Scalability, computation efficiency, compatibility, innovation, and interoperability are the top five factors that can highly influence CC adoption in the organization. Practitioners and managers can focus on these specific factors during the adoption of CC in order to attain effective deployment of cloud computing in I4.0-based production systems.

10.5 CONCLUSION

Cloud manufacturing adopts the cloud computing paradigm, which encompasses the idea of production as a service. Cloud computing offers several advantages, including infrastructure and cost savings. Industry practitioners are focused on implementing Industry 4.0-based advanced production systems in the industry. Cloud computing technology plays a significant role during the deployment of Industry 4.0 in the organization. In order to adopt cloud computing in the industry, practitioners must recognize factors that influence CC implementation. To fulfill this, the chapter focused on identifying 18 factors influencing the adoption of CC and evaluated them using a fuzzy TOPSIS approach. Fuzzy TOPSIS enabled ranking among factors that influence the implementation of CC in I4.0-based modern production systems. Scalability, computation efficiency, compatibility, innovation, and interoperability are the top five factors that can highly influence cloud computing adoption in the organization. This chapter provides insights for practitioners and managers in order to attain the successive implementation of cloud computing in the industry. In the future, additional relevant factors could be focused. Also, relevant solutions based on prioritized factors could be deployed.

REFERENCES

[1]. Malaga, A., & Vinodh, S. (2021). Benchmarking smart manufacturing drivers using grey TOPSIS and COPRAS-G approaches. *Benchmarking: An International Journal*, Vol. 28 No. 10, pp. 2916–2951. doi:10.1108/BIJ-12-2020-0620

[2]. Askary, Z., & Kumar, R. (2020). Cloud computing in industries: A review. *Recent Advances in Mechanical Engineering*, 107–116. doi:10.1007/978-981-15-1071-7_10

[3]. Badie, N., Hussin, A. R. C., & Lashkari, A. H. (2015). Cloud computing data center adoption factors validity by fuzzy AHP. *International Journal of Computational Intelligence Systems*, Vol. 8 No. 5, pp. 854–873.

[4]. Ali, O., Soar, J., Yong, J., & Tao, X. (2016, January). Factors to be considered in cloud computing adoption. In *Web Intelligence* (Vol. 14, No. 4, pp. 309–323). IOS Press. doi:10.3233/WEB-160347

[5]. Vater, J., Harscheidt, L., & Knoll, A. (2019, July). A reference architecture based on edge and cloud computing for smart manufacturing. In *2019 28th International Conference on Computer Communication and Networks (ICCCN)* (pp. 1–7). IEEE. doi:10.1109/ICCCN.2019.8846934

[6]. Hasan, L. M., Zgair, L. A., Ngotoye, A. A., Hussain, H. N., & Najmuldeen, C. (2015). A review of the factors that influence the adoption of cloud computing by small and medium enterprises. *Scholars Journal of Economics, Business and Management*, Vol. 2 No. 1, pp. 842–848.

[7]. Oliveira, T., Thomas, M., & Espadanal, M. (2014). Assessing the determinants of cloud computing adoption: An analysis of the manufacturing and services sectors. *Information & Management*, Vol. 51 No. 5, pp. 497–510.

[8]. Johnston, K. A., Loot, M., & Esterhuyse, M. P. (2016). The business value of cloud computing in South Africa. *The African Journal of Information Systems*, Vol. 8 No. 2, Article 1.

[9]. Jayasekara, D., Pawar, K., & Ratchev, S. (2019, June). A framework to assess readiness of firms for cloud manufacturing. In *2019 IEEE International Conference on Engineering, Technology and Innovation (ICE/ITMC)* (pp. 1–12). IEEE. doi:10.1109/ICE.2019.8792648

[10]. Liu, Y., Wang, L., Wang, X. V., Xu, X., & Jiang, P. (2019). Cloud manufacturing: Key issues and future perspectives. *International Journal of Computer Integrated Manufacturing*, Vol. 32 No. 9, pp. 858–874.

[11]. Misra, S. C., Rahi, S. B., Bisui, S., & Singh, A. (2019). Factors influencing the success of cloud adoption in the semiconductor industry. *Software Quality Professional*, Vol. 21 No. 2, pp. 38–51.

[12]. Alsafi, T., & Fan, I. S. (2020, June). Cloud computing adoption barriers faced by Saudi manufacturing SMEs. In *2020 15th Iberian Conference on Information Systems and Technologies (CISTI)* (pp. 1–6). IEEE. doi:10.23919/CISTI49556.2020.9140940

[13]. Vinodh, S., Thiagarajan, A., & Mulanjur, G. (2014). Lean concept selection using modified fuzzy TOPSIS: A case study. *International Journal of Services and Operations Management*, Vol. 18 No. 3, pp. 342–357.

[14]. Yoo, S. K., & Kim, B. Y. (2018). A decision-making model for adopting a cloud computing system. *Sustainability*, Vol. 10 No. 8, p. 2952. doi:10.3390/su10082952

[15]. Dobrescu, R., Mocanu, S., Chenaru, O., Nicolae, M., & Florea, G. (2021). Versatile edge gateway for improving manufacturing supply chain management via collaborative networks. *International Journal of Computer Integrated Manufacturing*, Vol. 34 No. 4, pp. 1–15. doi:10.1080/0951192X.2021.1879401

[16]. Wang, Y., Nekovee, M., Khatib, E. J., & Barco, R. (2021b). Machine learning/AI as IoT enablers. In *Wireless Networks and Industrial IoT* (pp. 207–223). Springer, Cham. doi:10.1007/978-3-030-51473-0_11

[17]. Anand, M. B., & Vinodh, S. (2018). Application of fuzzy AHP–TOPSIS for ranking additive manufacturing processes for microfabrication. *Rapid Prototyping Journal*, Vol. 24 No. 2, pp. 424–435. doi:10.1108/RPJ-10-2016-0160

11 Liberating 3D Printing for a New Normal in Manufacturing

Tarun Kataray, Mayank Mishra, Vaishnav Madhavadas,
Deepesh Padala, Bhawana Choudhary, Swati Kashyap,
Utkarsh Chadha, and S. Aravind Raj

CONTENTS

11.1 Introduction 103
11.2 How Additive Manufacturing Will Help Achieve a New Normal in Its
Varied Applications 104
  11.2.1 Aerospace Industry 105
  11.2.2 Automotive Industry 105
  11.2.3 Defense Industry 105
  11.2.4 Medical Industry 105
  11.2.5 Food Industry 106
  11.2.6 Fashion Industry 106
  11.2.7 Electric and Electronic Industry 106
  11.2.8 Art and Culture 106
  11.2.9 Construction Industry 107
11.3 Future Directions of AM/3D Printing for Industry 5.0 107
  11.3.1 Deep Learning and Machine Learning in 3D Printing Technology.................108
    11.3.1.1 Choosing Parameters...108
    11.3.1.2 Process Monitoring 108
  11.3.2 IOT in 3D Printing Technology 109
    11.3.2.1 IOT for Control and Monitoring of AM...109
    11.3.2.2 Beacon Technology 110
  11.3.3 Blockchain in 3D Printing Technology 110
    11.3.3.1 Checking Authenticity..110
    11.3.3.2 Protection of Intellectual Property 111
    11.3.3.3 Transparent, Secure, and Paperless Manufacturing 111
11.4 Conclusions 111
References 112

11.1 INTRODUCTION

Rapid prototyping, established in the 1980s for making models and prototype components, was the first method of creating a three-dimensional object layer by layer using computer-aided design (CAD). This technology was developed to aid in implementing engineers' vision. One of the first additive manufacturing (AM) technologies was rapid prototyping. It enables the development of printed parts as well as models. Among the primary breakthroughs this approach brought to product creation were time and cost savings, human contact, the product development cycle [1], and the ability to make practically any form previously impossible to machine. Rapid prototype methods are being utilized for more than simply generating models; with advances in plastic materials, it is now feasible to build whole things—of course, they were first designed to broaden the conditions examined in the prototyping process [2].

DOI: 10.1201/9781003186670-11

Rapid prototyping is not always the ideal choice; computer numerical control (CNC) machining procedures must be employed in some circumstances. The part dimensions may be bigger than currently available additive manufacturing printers [3]. Rapid prototyping materials are still scarce. Printing metals and ceramics are doable, but not all are routinely used industrial materials [4]. Over the last 30 years, the contribution and value of AM to advance manufacturing practice has altered dramatically. Initially designed to create prototypes for new items, its applicability has expanded to tooling and, more recently, direct manufacturing of end-use components or entire products as technologies have progressed. The overarching research focuses on individual manufacturing methods, one component of AM that has remained constant throughout its growth. Much focus has been placed on the prospects provided by AM machines, but this is sometimes at the expense of other key components of the production system. While there is no question that AM machines have numerous unique capabilities, assuming that they accomplish this alone in real-world production is an oversimplification. Various resources assist and complement AM in practice, although their contribution is rarely recognized in research. The idea that one can "simply press print" to produce does not match existing practice, and such exaggeration of technology capabilities has the potential to disenfranchise potential AM users.

This chapter gives an altogether new approach and investigation addressing the liberation of 3D printing for a new normal in manufacturing. The study technique includes searching for terms such as "AM," "Additive Manufacturing," "Machine Learning in Additive Manufacturing," "Blockchain in Additive Manufacturing," and "IOT in Additive Manufacturing" on SCOPUS, Google Scholar, Web of Science, and ScienceDirect. Two authors assessed the papers separately, matching the titles and abstracts to the inclusion criteria for potentially suitable publications. According to the citation and content field, the authors read more than 100 publications. Finally, 67 articles were chosen. This selection is based on the relevance and value of the study field for further investigation. This chapter presents the findings of existing publications from reputable journals through careful reviews.

Further study was conducted on many research topics and their implementation by finding, choosing, and analyzing knowledge pertinent understanding the problem. This chapter is appropriate for attaining the overall aims of the study, generalizing, and providing suggestions based on the findings. Numerous connected journals, papers, and books on sustainability, the environment, and other similar issues were read to perform this literature study.

11.2   HOW ADDITIVE MANUFACTURING WILL HELP ACHIEVE A NEW NORMAL IN ITS VARIED APPLICATIONS

Additive manufacturing is one of the newest additions to manufacturing processes. Additive manufacturing can make very simple designs and complex shapes, thereby making it more versatile and efficient than other manufacturing techniques. Additive manufacturing, due to less material wastage, lag time, and cost of operation compared to conventional manufacturing techniques, is a suitable alternative. As technological development occurs, most industries are starting to incorporate additive manufacturing for manufacturing components. One of the easiest ways to increase the efficiency of a component has been found to be by increasing the complexity of the component, and since it is easy to manufacture components with complex structures using AM, it is preferred over conventional manufacturing techniques.

In this section, the different applications of AM will be discussed. Due to the versatile nature of AM, it can be used to manufacture most products made from various materials. AM can be used for metals, metal alloys, polymers, ceramics, composites, and many more such materials. So for these reasons, AM is being used in many industries such as the aerospace industry, automotive industry, defense industry, medical industry, food industry, fashion industry, electric and electronic industry, art and culture, and construction industry.

11.2.1 Aerospace Industry

In the aerospace industry, metal AM is used to make complex structures and structures that require a very high strength-to-weight ratio. In commercial airplanes, it is necessary to reduce the amount of fuel used, and using components with a high strength-to-weight ratio will be helpful to solve this problem, thereby reducing the overall cost of flying the airplane [5]. The main metal AM methods used for manufacturing the metal components are powder bed fusion (PBF) and directed energy deposition (DED). PBF is used to manufacture complex components that are small in size and require high precision to make, whereas if the component is less complex and bigger, DED can be used. PBF is normally used to make components like blades of turbines and complicated engine components. DED is used to make wing ribs, flanges, and stiffened panels. Fused deposition modelling (FDM) is used to make prototypes of aerospace components using polymers. Selective laser sintering (SLS) and selective laser melting (SLM) are used to make prototypes of aerospace components using metals or metal alloys [6–10]. The method and material used to make prototypes are made based on the function of the actual component. Materials such as titanium grade 5 and Inconel can also be used in AM, making it suitable for components that require very high strength-to-weight ratio, corrosion resistance, heat resistance, and durability [11–16]. 3D printers can be used in space shuttles to make components in case of failure of a certain component, which is much more cost efficient than bringing the space shuttle back to earth, fixing it, and then launching it again.

11.2.2 Automotive Industry

AM is being used extensively in the automotive industry. The main reason for this is that AM can be used for weight reduction of an automotive component. Other reasons for this include the complexity of the components, the capability to work at high temperatures, moisture resistance, and part consolidation. Most automotive components can be 3D printed, and the main drawbacks of 3D printing are that mass production of the components is not cost efficient, and the build size is smaller compared to conventional manufacturing (CM) [17, 18]. It was found that the engine of light distribution trucks made using 3D printing was much more environmentally friendly than those made using CM methods. AM is also used to make components in luxury vehicles made in low volumes. AM is also used for motorsports, as it requires a very high strength-to-weight ratio, and the complexity of the component is also very high [19].

11.2.3 Defense Industry

Additive manufacturing is being used extensively in the defense sector. It is important to make complex components cheaply in the defense sector. Due to the uncertainty in the environment, there can be disruptions in the supply of weapons, ammo, and much more crucial equipment. AM is a suitable solution in such situations, as all the equipment can be 3D printed from any location with the design file and the raw materials. It is easier to transport raw materials than to transport the equipment itself. AM is mainly used to make drones and disposable medical supplies [20]. Even in 3D printing of these components, it is found that the use of wire feed is better than powder feed in terms of the deposition rate and the cost efficiency. Wire arc additive manufacturing (WAAM) is also used to manufacture defense equipment [21].

11.2.4 Medical Industry

Previously, AM was only used to make cells, organs, and implant models. However, AM is being used extensively in the medical industry due to advancement in AM technology. The development of bio-ink has helped to use AM in the medical industry to a very high extent. AM is currently used for drug delivery systems, medical equipment, orthopedics and scaffold formation, dentistry, and prosthetics. FDM is mainly used for 3D printed components in the medical industry. The main materials used are polymers, composites, and metals based on the application.

11.2.5 FOOD INDUSTRY

Food 3D printing is a relatively new 3D printing technology. The type of 3D printing used to print food is called food layer manufacturing (FLM). At present, AM cannot be used to manufacture food products. AM in the food industry is mainly used to change the shape and incorporate the required nourishment. The different AM processes used in the food industry are stereolithographic apparatus (SLA), binder jetting, selective laser sintering (SLS), and fused deposition modelling (FDM). The main materials used in the food industry are food polymer powder, fat-based compounds, sugar, starch, corn flour, flavors, and a fluid cover. Recently it was found that even meat can be 3D printed using certain biological ink. Therefore, it would be possible to have meat without killing animals in the future.

11.2.6 FASHION INDUSTRY

Additive manufacturing can be used in the textile industry and for making jewelry. Additive manufacturing can be used in textiles to enhance thermal conductivity, roughness, and porosity. It was found that washing significantly reduced the adhesion of the 3D printed material, so a pre-treatment was required. Adding rigid materials such as metal and polymer in fabric was beneficial. 3D printing can add polymers to increase the rigidity of the material or fabric used. This technology can be used to make orthopedic devices. 3D printing can also be used for printing elastic materials. Overall, 3D printing can make shoes, clothes, and gloves [22].

Most of the materials used to make jewelry are very costly. Due to this, it is important to keep the material wastage as low as possible; thereby, AM is the most suitable manufacturing method. Compared to garments, the material options for jewelry are much vaster in AM. Many direct and indirect methods of AM can be used to make jewelry. The main advantage of 3D-printed jewelry is that it is highly customizable and can be made for a cheaper price with the same quality or even higher quality than CM jewelry. Fashion designers still refrain from using 3D-printed jewelry, as it can easily be replicated using a high-end 3D printer, which does not make their product unique anymore, thereby reducing the market as well as the price of handmade jewelry, which is an advantage for customers but a loss for the jewelry company [23].

11.2.7 ELECTRIC AND ELECTRONIC INDUSTRY

AM can make active electronic components such as transistors, LEDs, organic thin-film transistor (OTFT), batteries, and so on. AM is mainly used in the electric and electronic industry due to its low material wastage. Due to the unsustainable use of the raw materials needed in the electric industry, the materials needed are very costly. Complex structures must make the component more efficient with less material. A lot of accuracy and precision are required to manufacture electrical and electronic components; therefore, AM is a very suitable method to manufacture these components. Due to less material wastage, even costlier and more efficient materials can be used to manufacture the components [24]. PBF is the most common method used for electrical and electronic component AM. The most commonly used materials for 3D printing of electrical components are metal, metal alloys, and polymers [25].

11.2.8 ART AND CULTURE

Art designers made customizable decorations possible using fabric (fabric paper) [22]. AM restored old Transylvanian popular pottery from the 16th–18th centuries [26] which would have been very difficult to restore using CM. AM is used nowadays to make personalized pots for a very cheap price. Pots with very complex designs can also be easily manufactured [27]. As technology advances, more designers and artists are moving towards AM to show their creativity, as it is much more cost efficient and has less lead time.

11.2.9 CONSTRUCTION INDUSTRY

In the construction industry, the use of AM can be extremely time saving and cheaper than the CM alternative. AM can shrink the supply chain, as the needed parts can be 3D printed on the site with a shorter lead time. Only the component design would be needed, reducing the price required to buy the finished product from other companies. The materials used mainly are cement, polymer, metals, and metal alloys. Material extrusion processes are the main AM processes used to manufacture cement components or large-scale applications [28]. The use of AM in the construction industry is 30% to 60% cheaper than using CM [29]. It has been found that manufacturing roofs, doors, windows, and so on is hard using AM [30].

As we talk about achieving the new normal, 3D printing technology has shown an aptitude for providing better digital solutions for day-to-day lives in this time of need [31]. The scope of digital technologies, however, is not restricted to VR, holography, and biosensing. In this new generation, present-day technologies have made possible the effective printing of a handful of exceptionally designed personal protection equipment (PPE) relevant to combat the COVID-19 pandemic, with less stress, time, and material usage. The sequential process involved in conventional testing challenges the death mean before the required number of mask production during the pandemic lockdown [32]. Some seriously affected countries controlled the death rate, such as China and other Asian nations, due to fast intervention through a complete stoppage by lockdown and mask use to forestall the spread of disease [33]. 3D printing in various industries like automotive, medical, electric, construction, and fashion helped satisfy demand, whereas some affected countries still faced some losses because of their slow action [34]. Except for in a few countries, the impact was less than expected because of AM technologies which majorly focused on 3D printing of materials for various applications, which may be majorly attributed to 3D printing materials [34].

The pandemic outbreak has given rise to studies in many areas. Within manufacturing, the urgent need for research covers a list of subjects, including the role 3D printing will play in managing COVID-19 and the effect of COVID-19 on 3D printing of equipment. After surveys and much research by scholars, we know that manufacturing is an important pandemic concern [35, 36]. Many countries have pointed out the purpose of manufacturing as one of the only ways for manufacturers to meet the sudden rise in demands for medical equipment like PPE and components for other industries [37]. 3D printing is playing a crucial role in guiding and building an eco-friendly economy, as technologies will strive to make the recovery from pandemics a sustainable and inclusive one. 3D printing has been made easy, as it consumes a lot less material than traditional manufacturing, as earlier the recycled part of the goods could have been returned to the particular production cycle [38]. In some ways, it has been incontestable, like employing a 3D printing initiative to avoid a lack of resources [39–41]. To supply solutions to the present state of affairs, there are carbon, form labs, and two organic compound 3D printer makers, and the World Health Organisation manufactures thousands of pieces of equipment for industries daily [42, 43].

11.3 FUTURE DIRECTIONS OF AM/3D PRINTING FOR INDUSTRY 5.0

Growing technologies have brought a new normal with improvised technologies of blockchain and additive manufacturing [44]. Industry 5.0 usually brings about a pattern that interrupts the way of operating for all factories [45, 46]. In addition to blockchain, it safeguards industrial properties and many other concerns. By improvising blockchain, the potential to authorize machine-to-machine connectivity in 3D can be seen in the future [47, 48]. Blockchain is gaining much attention from organizations, especially in the supply chain. When combined with the IoT and machine learning (ML), Blockchain helps make the supply chain leaner and far better than before [49, 50]. Thus, blockchain in the 3D printing supply chain has a promising future and can provide many solutions to current problems [51].

11.3.1 Deep Learning and Machine Learning in 3D Printing Technology

Machine learning is a field in computer science that deals with machines learning how to perform certain tasks, such as regression, classification, generation, and so on. This is accomplished using algorithms that mathematically model the relationship between various parameters that affect the assigned task. In the case of additive manufacturing, the parameters [52] include the material, infill density, extrusion temperature, raster angle, ultimate tensile strength, and so on. ML can study the relation between these parameters and perform regression or classification according to the chosen parameter, or it can correlate it with a different one that might be affected by these parameters.

While some of these relations can be modeled with sufficient knowledge on the subject matter (expert systems [53]), it is a very time-consuming process where ML comes in. The ML algorithms that learn these relations are based on statistics and probability theory. They have been observed to come up with very accurate approximations of the actual relation. These approximations do not have a 100% accuracy rate; however, they are generated in a very short amount of time and for much less computational power. This is the trade-off most scientists who apply ML algorithms make. These algorithms have also demonstrated that they can generate highly complex relations close to impossible for humans to notice.

As mentioned, ML algorithms never exhibit a 100% accuracy rate in complex tasks. These algorithms are simple and efficient but do not even get close to a 100% accuracy rate. This is where deep learning (DL) comes in handy. It is a subset of ML which comprises very complex models modeled after neural networks found in brains. Deep neural networks (DNNs) are based on the universal approximation theorem (UAT) [54]. The UAT states that a DNN can approximate any relation with sufficient accuracy, provided the neural network is deep enough.

A DNN consists of layers of neurons or perceptrons. These perceptrons take input parameters such as the ones mentioned previously (material, infill density, extrusion temperature, etc.) arranged in a vector and multiply them by a set of weights which are initialized randomly and then added to a bias. This is known as the feed-forward stage (as in Figure 11.1). The resulting integer is pushed forward to the activation stage, where nonlinearity is introduced to the system. In the feed-forward stage in Figure 11.1, the parameters are multiplied by randomly initialized weights and added to a randomly initialized bias to give an output. The weights and biases are updated to fit the data used to train the perceptron using the gradient descent algorithm. Multiple such perceptrons can be stacked to form layers, and multiple such layers can be stacked to form a deep neural network. In conjunction, the activation functions and the feed-forward stage can approximate any complex relationship between the input and the output, which could be any other parameter in the dataset, such as the ultimate tensile strength.

11.3.1.1 Choosing Parameters

There are a lot of process parameters [55] in the 3D printing process, as shown in Table 11.1. For a new adopter, this can be overwhelming and may cause defects that an experienced person might avoid. In order to make it easier for newcomers and aid experienced people, a DNN can be used to get the optimal process parameters [56] and the best possible results.

11.3.1.2 Process Monitoring

Another area where ML is highly used in AM is process monitoring [57]. Inputs from various sensors, such as cameras and thermal sensors, can train neural networks based on the quality of the prints.

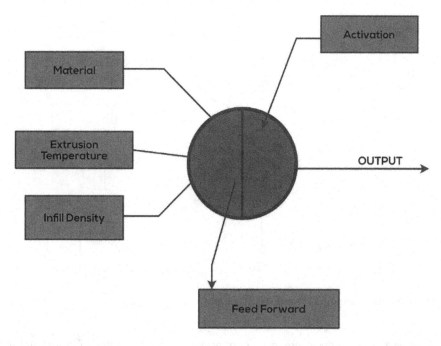

FIGURE 11.1 Overview of a perceptron.

TABLE 11.1

Examples of Variables That Can Be Collected to Train ML Algorithms

Process Parameters	Resultant Data	Microstructure Parameters	Mechanical Properties
Extrusion temperature	Printing time	Pores	Ultimate tensile strength (UTS)
Raster angle Material	Cost	Surface smoothness	Young's modulus
Filament size	Sounds	Grain size	
Nozzle size			
Bed temperature			

As discussed in the next section, when used with cloud computing, this makes for a very user-friendly system, which may perform better than a human observing the printer.

11.3.2 IOT in 3D Printing Technology

IoT stands for the Internet of Things, a system of interconnected devices that communicate. These devices run on embedded systems, including processors, sensors, and communication hardware. Sensors collect data from the environment depending on the use case, which it then processes to attain insights that it can communicate with the rest of the devices in the mesh.

11.3.2.1 IOT for Control and Monitoring of AM

The IoT is a powerful technology harnessed to push additive manufacturing to the next level. There are many sensors in a 3D printer that generate a lot of data. Those data can be relayed to other mobile devices connected to the same cloud service, following an architecture similar to the one shown in Figure 11.2. A mobile app was developed [58] to display variables such as the estimated

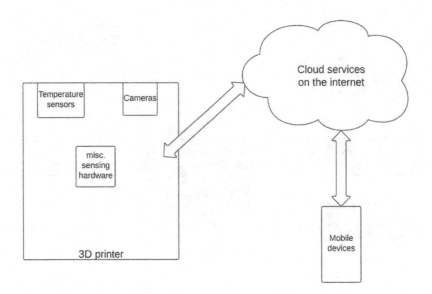

FIGURE 11.2 Architecture of the IoT-enabled 3D printer.

printing time, heated bed temperature, percent of printing volume, number of layers, and so on, all of which could be read through the sensing hardware mounted on it or the ML algorithms presented previously, in which the agent could be used to make decisions and control the printer by starting/ stopping/pausing the printer and controlling parameters such as the bed or extruder temperature, speed multiplier [58], and so on which could be used to get a better result.

11.3.2.2   Beacon Technology

The previous implementation had the 3D printer connected directly to the cloud, which made the implementation quite expensive. This is where beacons display their efficiency. 3D printers were equipped with beacons [59], transmitting signals to a local mobile device using Bluetooth technology. That information is then transmitted to the cloud through Wi-Fi to be parsed and processed to the user.

11.3.3   Blockchain in 3D Printing Technology

Blockchain is a relatively new technology that has made fingerprinting and recording tasks very easy and secure. It can be defined as a decentralized, public ledger across a network [60]. It has gained much popularity with the rise of crypto-currencies, but it can be used for so much more. This section will highlight the uses of blockchain technology in additive manufacturing.

In the previous section about IoT technology, there was a lot of exchange of information, including and not limited to stereolithography (STL) files [56], through various methods (Wi-Fi, internet, Bluetooth, etc.). While this makes it very convenient for the users and significantly improves the accuracy and customizability, it adds many vulnerabilities to the system. Recent studies have uncovered how blockchain can be used in AM to protect intellectual property or monetize AM for business models [61]. It can also make monitoring a 3D printer from a cloud service provider much more secure.

11.3.3.1   Checking Authenticity

AM is being adapted by many critical industries these days, which requires much security. Industries such as the medical and defense industry will face dire consequences if the data used to print products are corrupted or tampered with. Most of the time goes into inspection of CAD models to prevent such a disaster from happening, which researchers have suggested fixing by adding the STL file to a blockchain network [61]. Due to the immutable nature of blockchain [60], these data cannot be tampered with, making it a very effective and time-saving implementation to ensure the authenticity of the data.

11.3.3.2   Protection of Intellectual Property

Intellectual property (IP) theft is still a major issue in many industries. This can be prevented by tagging data that corresponds to the IP [62]. A very important property of these tags is that any attempt to tamper with them has to be futile. This is where blockchain can tag the data regarding the CAD designs. Adding data to a blockchain network ensures that they can never be tampered with. It provides a good platform for tagging data, protecting IP, and avoiding plagiarism.

11.3.3.3   Transparent, Secure, and Paperless Manufacturing

Transparency is a new addition that will make manufacturing and supply chains much more effective. Researchers have proposed a prototype known as "FabRec" [63, 64], a system of distributed, transparent, and verified manufacturing stations that is autonomous to a certain extent. The feasibility of a decentralized network of automated manufacturing stations was displayed very well in the prototype and can be improved to tackle many issues faced in the sector.

11.4   CONCLUSIONS

This chapter presents an overview of 3D printing technologies in the manufacturing industry. In addition, various applications of additive manufacturing are discussed in this chapter. These applications include aerospace, automotive, defense, medical, food, fashion, electric, and construction industries. This chapter also discusses future directions for additive manufacturing in the Industry 5.0 revolution [65, 66]. This chapter further discusses the role of new technologies like deep learning, machine learning, blockchain, and the IoT in additive manufacturing. Furthermore, the important conclusions of this chapter are presented in the following [67].

(i)   Without a single doubt, 3D printing technologies are a principal component in the next major industrial revolution. Due to its versatility, 3D printing plays a crucial role in Industry 5.0, saving time or cost of production. Presently, additional industrial segments are finding the advantages of AM. The advanced factories for the long term have all their processes interconnected by the IoT, incorporating greater flexibility in production processes.

(ii)   The latest advancements in information technology coalesce very well with AM. Recent advancements in AI/ML, the IoT, and blockchain can be used for

a) Reducing production costs,
b) Reducing errors,
c) Reducing waste,
d) Reducing production time,
e) Conserving energy,
f) Increasing security.

These goals are achieved by using ML techniques to present printing parameters to get the best possible outcome in the least possible time. Researchers have used the IoT to set up remote monitoring and control through the cloud.

(iii)   AM processes and the previously mentioned embellishments deal with much of the data exchanged, making the system prone to cyber-attacks. That can be prevented using blockchain, with fingerprinting methods for STL files to prevent IP theft and plagiarism and ways to verify and control the other processes securely.

   (iv) Due to the versatile nature of AM, it can be used in most industries for small-scale production. Based on the application of the component, the AM process used, feed mechanism, lead time, accuracy, and so on change. It is found that wire feed is the best feeding mechanism for getting accurate prints. Materials such as titanium grade 5, Inconel, and many more costly materials can be used in AM due to their minimum material wastage.

   (v) As technology advances, new materials such as proteins, fat-based compounds, and so on are also being used in AM. Within a few years, AM may be feasible for large-scale manufacturing.

REFERENCES

[1] Wohlers T. Rapid prototyping systems. In *Proceedings of the First Rapid Prototyping Convention*, Paris, France, June 1993.

[2] Cooper K. Rapid prototyping technology. *Rapid Prototyping Technology*, 2001. https://doi. org/10.1201/9780203910795/RAPID-PROTOTYPING-TECHNOLOGY-KENNETH-COOPER.

[3] Wong KV, Hernandez A. A review of additive manufacturing. *International Scholarly Research Notices* 2012; Article ID 208760, 10 pages. https://doi.org/10.5402/2012/208760

[4] Kruth JP. Material incress manufacturing by rapid prototyping techniques. *CIRP Annals*, 1991;40(2):603–614.

[5] Gisario A, Kazarian M, Martina F, Mehrpouya M. Metal additive manufacturing in the commercial aviation industry: A review. *Journal of Manufacturing Systems* 2019;53:124–149.

[6] Ngo TD, Kashani A, Imbalzano G, Nguyen KT, Hui D. Additive manufacturing (3D printing): A review of materials, methods, applications and challenges. *Composites Part B: Engineering* 2018;143:172–196.

[7] Katz-Demyanetz A, Popov Jr VV, Kovalevsky A, Safranchik D, Koptioug A. Powder-bed additive manufacturing for aerospace application: Techniques, metallic and metal/ceramic composite materials and trends. *Manufacturing Review* 2019;6(5).

[8] Bhavar V, Kattire P, Patil V, Khot S, Gujar K, Singh R. A review on powder bed fusion technology of metal additive manufacturing. *Additive Manufacturing Handbook*, 2018: 251–3. https://doi. org/10.1201/9781315119106-15/REVIEW-POWDER-BED-FUSION-TECHNOLOGY-METAL-ADDITIVE-MANUFACTURING-VALMIK-BHAVAR-PRAKASH-KATTIRE-VINAYKUMAR-PATIL-SHREYANS-KHOT-KIRAN-GUJAR-RAJKUMAR-SINGH.

[9] Ahn DG. Direct metal additive manufacturing processes and their sustainable applications for green technology: A review. *International Journal of Precision Engineering and Manufacturing—Green Technology* 2016;3:381–395. https://doi.org/10.1007/S40684-016-0048-9.

[10] Manfredi D, Calignano F, Ambrosio EP, Krishnan M, Canali R, Biamino S, et al. Direct metal laser sintering: An additive manufacturing technology ready to produce lightweight structural parts for robotic applications. *FracturaeCom* 2013;10.

[11] Kazantseva N. Main factors affecting the structure and properties of titanium and cobalt alloys manufactured by the 3D printing. *Journal of Physics: Conference Series* 2018;1115(4):042008.

[12] Karolewska K, Ligaj B. Comparison analysis of titanium alloy Ti6Al4V produced by metallurgical and 3D printing method. In *AIP Conference Proceedings* 2019;2077(1):020025.

[13] Merriam EG, Jones JE, Howell LL. Design of 3D-printed titanium compliant mechanisms. In *The 42nd Aerospace Mechanism Symposium*, May, 2014.

[14] Chen W, Li Z. Additive manufacturing of titanium aluminides. In *Additive Manufacturing for the Aerospace Industry* (pp. 235–263). Elsevier, 2019.

[15] MacDonald D, Fernández R, Delloro F, Jodoin B. Cold spraying of Armstrong process titanium powder for additive manufacturing. *Journal of Thermal Spray Technology* 2017;26:598–609. https://doi. org/10.1007/S11666-016-0489-2.

[16] Uhlmann E, Kersting R, Klein TB, Cruz MF, Borille AV. Additive manufacturing of titanium alloy for aircraft components. *Procedia Cirp* 2015;35:55–60.

[17] Prakash KS, Nancharaih T, Rao VVS. Additive Manufacturing Techniques in Manufacturing—An Overview. *Materials Today: Proceedings* 2018;5:3873–3882. https://doi.org/10.1016/J.MATPR.2017.11.642.

[18] Sarvankar SG, Yewale SN. Additive manufacturing in automobile industry. International *Journal of Research in Aeronautical and Mechanical Engineering* 2019;7(4):1–10.

[19] Böckin D, Tillman AM. Environmental assessment of additive manufacturing in the automotive industry. *Journal of Cleaner Production* 2019;226:977–987. https://doi.org/10.1016/J.JCLEPRO.2019.04.086.

[20] Busachi A, Erkoyuncu J, Colegrove P, Martina F, Watts C, Drake R. A review of additive manufacturing technology and cost estimation techniques for the defence sector. *CIRP Journal of Manufacturing Science and Technology* 2017;19:117–128. https://doi.org/10.1016/j.cirpj.2017.07.001.

[21] Busachi A, Erkoyuncu J, Colegrove P, Martina F, Ding J. Designing a WAAM based manufacturing system for defence applications. *Procedia CIRP* 2015;37:48–53. https://doi.org/10.1016/J.PROCIR.2015.08.085.

[22] Sitotaw DB, Ahrendt D, Kyosev Y, Kabish AK. Additive manufacturing and textiles—state-of-the-art. *Applied Sciences* 2020;10(15):5033.

[23] Yap YL, Yeong WY. Additive manufacture of fashion and jewellery products: A mini review. *Virtual and Physical Prototyping* 2014;9:195–201. https://doi.org/10.1080/17452759.2014.938993.

[24] Saengchairat N, Tran T, Chua CK. A review: Additive manufacturing for active electronic components. *Virtual and Physical Prototyping* 2017;12(1):31–46. https://doi.org/10.1080/17452759.2016.1253181.

[25] Wrobel R, Mecrow B. A comprehensive review of additive manufacturing in construction of electrical machines. *IEEE Transactions on Energy Conversion* 2020;35:1054–1064. https://doi.org/10.1109/TEC.2020.2964942.

[26] Măruțoiu C, Bratu I, Țiplic MI, Măruțoiu VC, Nemeș OF, Neamțu C, Hernanz, A. FTIR analysis and 3D restoration of Transylvanian popular pottery from the XVI-XVIII centuries. *Journal of Archaeological Science: Reports* 2018;19:148–154.

[27] Cai R, Lin Y, Li H, Zhu Y, Tang X, Weng Y, et al. Wowtao: A personalized pottery-making system. *Computers in Industry* 2021;124. https://doi.org/10.1016/J.COMPIND.2020.103325.

[28] Paolini A, Kollmannsberger S, Rank E. Additive manufacturing in construction: A review on processes, applications, and digital planning methods. *Additive Manufacturing* 2019;30:100894.

[29] Shakor P, Nejadi S, Paul G, Malek S. Review of emerging additive manufacturing technologies in 3d Printing of cementitious materials in the construction industry. *Frontiers in Built Environment* 2019;4. https://doi.org/10.3389/FBUIL.2018.00085/FULL.

[30] Craveiroa F, Duartec JP, Bartoloa H, Bartolod PJ. Additive manufacturing as an enabling technology for digital construction: A perspective on Construction 4.0. *Automation in Construction* 2019;103:251–267.

[31] Yan Q, Dong H, Su J, Han J, Song B, Wei Q, Shi Y. A review of 3D printing technology for medical applications. *Engineering* 2018;4(5):729–742.

[32] Dodziuk H. Application of 3D printing in healthcare of the elderly. *Gerontologia Polska* 2019;27:293–299.

[33] Jammalamadaka U, Tappa K. Recent advances in biomaterials for 3D printing and tissue engineering. *Journal of Functional Biomaterials* 2018;9(1):22.

[34] Bandyopadhyay A, Bose S, Das S. 3D printing of biomaterials. *MRS Bulletin* 2015;40(2):108–115.

[35] Prather KA, Wang CC, Schooley RT. Reducing transmission of SARS-CoV-2. *Science* 2020;368(6498):1422–1424.

[36] Kis Z, Kontoravdi C, Dey AK, Shattock R, Shah N. Rapid development and deployment of high-volume vaccines for pandemic response. *Wiley Online Library* 2020;2. https://doi.org/10.1002/amp2.10060.

[37] López-Gómez C, Corsini L, Leal-Ayala D, Fokeer S. COVID-19 critical supplies: The manufacturing repurposing challenge. United Nations Industrial Development Organization, 2020.

[38] Advincula R, Ryan Dizon JC, Niu I, Chung J, Kilpatrick L, Newman R. Additive manufacturing for COVID-19: Devices, materials, prospects, and challenges. *Cambridge* 2020;10:413–427. https://doi.org/10.1557/mrc.2020.57.

[39] De Leon AC, Chen Q, Palaganas NB, Palaganas JO, Manapat J, Advincula RC. High performance polymer nanocomposites for additive manufacturing applications. *Reactive and Functional Polymers* 2016;103:141–155.

[40] Valino AD, Dizon JRC, Espera Jr AH, Chen Q, Messman J, Advincula RC. Advances in 3D printing of thermoplastic polymer composites and nanocomposites. *Progress in Polymer Science* 2019;98:101162.

[41] Manapat JZ, Chen Q, Ye P, Advincula RC. 3D printing of polymer nanocomposites via stereolithography. *Macromolecular Materials and Engineering* 2017;302. https://doi.org/10.1002/MAME.201600553.

[42] Dizon JRC, Espera Jr AH, Chen Q, Advincula RC. Mechanical characterization of 3D-printed polymers. *Additive Manufacturing* 2018;20:44-67.

[43] Roberge RJ. Face shields for infection control: A review. *Journal of Occupational and Environmental Hygiene* 2016;13:239–246. https://doi.org/10.1080/15459624.2015.1095302.

[44] Holland M, Stjepandić J, Nigischer C. Intellectual property protection of 3D print supply chain with blockchain technology. In *2018 IEEE International Conference on Engineering, Technology and Innovation (ICE/ITMC)* (pp. 1–8). IEEE, June 2018.

[45] Guo D, Ling S, Li H, Ao D, Zhang T, Rong Y, Huang GQ. A framework for personalized production based on digital twin, blockchain and additive manufacturing in the context of Industry 4.0. In *2020 IEEE 16th International Conference on Automation Science and Engineering (CASE)* (pp. 1181–1186). IEEE, August, 2020.

[46] Chen F, Mac G, Gupta N. Security features embedded in computer aided design (CAD) solid models for additive manufacturing. *Materials & Design* 2017;128:182–194.

[47] Chen Q, Han L, Ren J, Rong L, Cao P, Advincula RC. 4D printing via an unconventional fused deposition modeling route to high-performance thermosets. *ACS Applied Materials and Interfaces* 2020;12:50052–50060. https://doi.org/10.1021/ACSAMI.0C13976.

[48] Wu D, Ren A, Zhang W, Fan F, Liu P, Fu X, Terpenny J. Cybersecurity for digital manufacturing. *Journal of Manufacturing Systems* 2018;48:3–12.

[49] Bridges SM, Keiser K, Sissom N, Graves SJ. Cyber security for additive manufacturing. *ACM International Conference Proceeding Series* 2015;06–08-April-2015. https://doi.org/10.1145/2746266.2746280.

[50] Yampolskiy M, Andel TR, McDonald JT, Glisson WB, Yasinsac A. Intellectual property protection in Additive Layer Manufacturing: Requirements for secure outsourcing. *ACM International Conference Proceeding Series 2014*;12-December-2014. https://doi.org/10.1145/2689702.2689709.

[51] Shahbazi Z, Byun YC. Smart manufacturing real-time analysis based on blockchain and machine learning approaches. *Applied Sciences* 2021;11(8):3535.

[52] Leite M, Fernandes J, Deus A, . . . LR-. . . conference on progress, 2018 undefined. Study of the influence of 3D printing parameters on the mechanical properties of PLA. *DrNtuEduSg* 2018:547–552. https://doi.org/10.25341/D4988C.

[53] Jackson P. Introduction to expert systems. United States. https://www.osti.gov/biblio/5675197

[54] Baker MR, Patil RB. Universal approximation theorem for interval neural networks. *Reliable Computing* 1998;10:235–239. https://doi.org/10.1023/A:1009951412412.

[55] Prabhakar MM, Saravanan AK, Lenin AH, Mayandi K, Ramalingam PS. A short review on 3D printing methods, process parameters and materials. *Materials Today: Proceedings* 2021;45:6108–6114.

[56] Wang Y, Lin Y, Zhong RY, Xu X. IoT-enabled cloud-based additive manufacturing platform to support rapid product development. *International Journal of Production Research* 2019;57(12):3975–3991.

[57] Amini M, Chang SI, Rao P. A cybermanufacturing and AI framework for laser powder bed fusion (LPBF) additive manufacturing process. *Manufacturing Letters* 2019;21:41–44.

[58] Barbosa GF, Aroca RV. An IoT-based solution for control and monitoring of additive manufacturing processes. *Journal of Powder Metallurgy & Mining* 2017;6(158):2.

[59] Ashima R, Haleem A, Bahl S, Javaid M, Mahla SK, Singh S. Automation and manufacturing of smart materials in Additive Manufacturing technologies using Internet of Things towards the adoption of Industry 4.0. *Materials Today: Proceedings* 2021;45:5081–5088.

[60] Nofer M, Gomber P, Hinz O, Schiereck D. Blockchain. *Business & Information Systems Engineering* 2017;59(3):183–187.

[61] Ghimire T, Joshi A, Sen S, Kapruan C, Chadha U, Selvaraj SK. Blockchain in additive manufacturing processes: Recent trends & its future possibilities. *Materials Today: Proceedings*, 2021.

[62] Sola A, Sai Y, Trinchi A, Chu C, Shen S, Chen S. How can we provide additively manufactured parts with a fingerprint? A review of tagging strategies in additive manufacturing. *Materials* 2021;15(1):85.

[63] Alkaabi N, Salah K, Jayaraman R, Arshad J, Omar M. Blockchain-based traceability and management for additive manufacturing. *IEEE Access* 2020;8:188363–188377.

[64] Angrish A, Craver B, Hasan M, Starly B. A case study for blockchain in manufacturing:"FabRec": A prototype for peer-to-peer network of manufacturing nodes. *Procedia Manufacturing* 2018;26:1180–1192.

[65] Chadha U, Abrol A, Vora NP, et al. Performance evaluation of 3D printing technologies: A review, recent advances, current challenges, and future directions. *Progress in Additive Manufacturing* 2022. https://doi.org/10.1007/s40964-021-00257-4

[66] Madhavadas V, Srivastava D, Chadha U, Raj SA, Sultan MTH, Shahar FS, Shah AUM. A review on metal additive manufacturing for intricately shaped aerospace components. *CIRP Journal of Manufacturing Science and Technology* 2022;39:18–36.

[67] Sachdeva I, Ramesh S, Chadha U, Punugoti H, Selvaraj SK. Computational AI models in VAT photopolymerization: A review, current trends, open issues, and future opportunities. *Neural Computing and Applications* 2022;1–23.

Index

0-9

3D Printing, 104, 105, 106, 107, 108, 111, 112, 113, 114

A

additive manufacturing, 34, 103, 104, 105, 106, 107, 108, 110, 111, 112, 113, 114
advanced analytics, 68
aerospace, 104, 105, 111, 112, 114
agility, 91
algorithms, 10, 11, 12, 13, 14, 17
artificial intelligence (AI), 15, 19, 23, 24, 27, 31, 57, 62, 68
AR/VR, 340
assembly design, 33
augmented reality, 37–45
autonomous robots, 10–17
autonomy, 10, 17

B

Bayes' filter, 12
big data, 19–28
    analytics, 31
    in industry, 19–28
blockchain, 104, 107, 108, 110, 111, 114

C

circularity, 3, 5, 6
clinical trial, 60, 62
cloud analyst, 50, 55
cloud based manufacturing, 20
cloud computing, 50, 51, 92
cloud manufacturing, 49, 50, 92, 93, 94, 100, 101
cloud production, 92
cloud service, 50, 51
cloud technology, 91, 93
collaborative robots, 18
compatibility, 92, 95, 100, 101
construction research, 89
COVID-19, 15
customers, 3, 4, 6
customization, 95
cyber physical systems, 31, 65, 67, 69

D

data collection, 66, 69
decentralized data centre, 51, 52
decision making, 94, 101
deep learning, 68
defense, 104, 105, 110, 111
design, 37, 38, 39, 41, 42, 46
digital, 19–28
    enablers, 21
    security, 35

supply chain, 3, 4
transformation, 2, 4, 6, 7
twin, 31–35
digitization, 19–28
directed-energy deposition (DED), 105
drug discovery, 60, 62

E

epidemic/pandemic outbreak prediction, 600

F

flexible, 65–72
flexible manufacturing, 20
food industry, 104, 106
food layer manufacturing (FLM), 106
fuzzy TOPSIS, 94, 101

I

implementation, 5, 6
Industrial Internet of Things (IIoT), 20
Industrial Revolution, 2, 4, 31
industrial robots, 10, 14
Industry 4.0, 1–6, 17, 19–28, 31, 34, 35, 37, 38, 40, 41, 66, 67, 76, 77, 80, 86, 87, 88 , 91, 92, 93, 94, 99, 100, 101
innovation, 91, 100, 101
intelligent machining, 31
Internet of Things (IoT), 2, 3, 5, 109, 111, 20, 21, 27, 69, 77, 88, 89

K

Kalman filter-based algorithm, 12

L

layout optimization, 32
logistics, 38
    with IoT, 19–28

M

machine learning, 57–62, 68, 69 , 104, 108, 111
machine to machine, 19, 20, 25
maintenance, 37–45
maintenance management, 65, 66, 67, 69
managing inventory, 62
manufacturing, 1–7, 37–38, 49, 50, 54, 55
manufacturing systems, 37, 38, 41, 92, 94, 99, 100
mapping, 17
medical, 104, 105, 106, 107, 110, 111, 113
mobile technology, 82
monitoring, 4
Monte Carlo localization, 13, 17

N

navigation, 9, 10, 11, 12, 13, 15, 17

P

path planning, 10, 13, 14, 17, 18
powder bed fusion (PBF), 105, 112, 114
predictive maintenance, 66, 71
product and process development, 38
product development, 3, 4, 6
production planning and control, 65, 69
productivity, 76, 87, 88
product life cycle, 66

Q

quality, 38, 41, 42, 43

R

real-time monitoring, 31–34
regional distributions, 52
reinforcement learning, 61
reliability, 80
repurposing, 60
roadmap, 5, 6, 7
robots, 9, 10, 12, 14, 15, 16, 17, 18

S

safety monitoring, 77
safety performance, 77, 87, 88, 89
scalability, 35

security, 92, 95
security monitoring, 77
selective laser melting (SLM), 105
selective laser sintering (SLS), 105, 106
semi-supervised learning, 57, 60
sensor(s), 10, 11, 15, 16, 76, 77, 78, 80, 83, 86, 87, 88, 89
simulation, 31–35, 49–55
simultaneous localization and mapping (SLAM), 11, 15, 17
smart factories, 76
smart manufacturing, 20, 33, 34, 35, 66, 69, 72
smart operators, 38
smart technology, 19–28
standardization, 35
supply chain, 1–7
supply chain management (SCM), 58, 61

U

unit systems, 71
unsupervised learning, 59, 60, 63
user bases, 51, 52, 53, 55

V

virtual manufacturing, 39, 41
virtual monitoring IoT, 31, 33, 34
virtual reality, 37–46

W

warehouse automation, 22, 24
wearable devices, 76
wire arc additive manufacturing (WAAM), 105